Standardisation in cell and tissue engineering

© Woodhead Publishing Limited, 2013

Related titles:

Progenitor and stem cell technologies and therapies
(ISBN 978-1-84569-984-0)

Electrospinning for tissue regeneration
(ISBN 978-1-84569-741-9)

Biomaterials and regenerative medicine in ophthalmology
(ISBN 978-1-84569-443-2)

Details of these books and a complete list of titles from Woodhead Publishing can be obtained by:

- visiting our web site at www.woodheadpublishing.com
- contacting Customer Services (e-mail: sales@woodheadpublishing.com; fax: +44 (0) 1223 832819; tel.: +44 (0) 1223 499140 ext. 130; address: Woodhead Publishing Limited, 80, High Street, Sawston, Cambridge CB22 3HJ, UK)
- in North America, contacting our US office (e-mail: usmarketing@ woodheadpublishing.com; tel.: (215) 928 9112; address: Woodhead Publishing, 1518 Walnut Street, Suite 1100, Philadelphia, PA 19102-3406, USA)

If you would like e-versions of our content, please visit our online platform: www. woodheadpublishingonline.com. Please recommend it to your librarian so that everyone in your institution can benefit from the wealth of content on the site.

We are always happy to receive suggestions for new books from potential editors. To enquire about contributing to our Biomaterials series, please send your name, contact address and details of the topic/s you are interested in to laura.overend@ woodheadpublishing.com. We look forward to hearing from you.

The team responsible for publishing this book:

Commissioning Editor: Laura Overend
Publications Coordinator: Emily Cole
Project Editor: Kate Hardcastle
Editorial and Production Manager: Mary Campbell
Production Editor: Mandy Kingsmill
Project Manager: Newgen Knowledge Works Pvt Ltd
Copyeditor: Newgen Knowledge Works Pvt Ltd
Proofreader: Newgen Knowledge Works Pvt Ltd
Cover Designer: Terry Callanan

© Woodhead Publishing Limited, 2013

Woodhead Publishing Series in Biomaterials: Number 58

Standardisation in cell and tissue engineering

Methods and protocols

Edited by
Vehid Salih

Oxford   Cambridge   Philadelphia   New Delhi

© Woodhead Publishing Limited, 2013

Published by Woodhead Publishing Limited,
80 High Street, Sawston, Cambridge CB22 3HJ, UK
www.woodheadpublishing.com
www.woodheadpublishingonline.com

Woodhead Publishing, 1518 Walnut Street, Suite 1100, Philadelphia,
PA 19102-3406, USA

Woodhead Publishing India Private Limited, 303, Vardaan House, 7/28 Ansari Road,
Daryaganj, New Delhi - 110002, India
www.woodheadpublishingindia.com

First published 2013, Woodhead Publishing Limited
© Woodhead Publishing Limited, 2013. Note: the publisher has made every effort to
ensure that permission for copyright material has been obtained by authors wishing to
use such material. The authors and the publisher will be glad to hear from any copyright
holder it has not been possible to contact.
The authors have asserted their moral rights.

This book contains information obtained from authentic and highly regarded
sources. Reprinted material is quoted with permission, and sources are indicated.
Reasonable efforts have been made to publish reliable data and information, but
the authors and the publisher cannot assume responsibility for the validity of all
materials. Neither the authors nor the publisher, nor anyone else associated with
this publication, shall be liable for any loss, damage or liability directly or indirectly
caused or alleged to be caused by this book.
    Neither this book nor any part may be reproduced or transmitted in any form or
by any means, electronic or mechanical, including photocopying, microfilming and
recording, or by any information storage or retrieval system, without permission in
writing from Woodhead Publishing Limited.
    The consent of Woodhead Publishing Limited does not extend to copying for
general distribution, for promotion, for creating new works, or for resale. Specific
permission must be obtained in writing from Woodhead Publishing Limited for
such copying.

Trademark notice: Product or corporate names may be trademarks or registered
trademarks, and are used only for identification and explanation, without intent to
infringe.

British Library Cataloguing in Publication Data
A catalogue record for this book is available from the British Library.

Library of Congress Control Number: 2013940961

ISBN 978-0-85709-419-3 (print)
ISBN 978-0-85709-872-6 (online)
ISSN 2049-9485 Woodhead Publishing Series in Biomaterials (print)
ISSN 2049-9493 Woodhead Publishing Series in Biomaterials (online)

The publisher's policy is to use permanent paper from mills that operate a
sustainable forestry policy, and which has been manufactured from pulp which is
processed using acid-free and elemental chlorine-free practices. Furthermore, the
publisher ensures that the text paper and cover board used have met acceptable
environmental accreditation standards.

Typeset by Newgen Knowledge Works Pvt Ltd, India
Printed by Lightning Source

© Woodhead Publishing Limited, 2013

Contents

	Contributor contact details	*xi*
	Woodhead Publishing Series in Biomaterials	*xv*
	Foreword	*xix*
	Introduction	*xxiii*

Part I	**Methods for cell and tissue engineering**	**1**

1	**Fundamentals of cell and matrix biology for tissue engineering**	**3**
	V. SALIH, Plymouth University Schools of Medicine and Dentistry, UK (formerly at UCL Eastman Dental Institute, UK) and D. THOMAS, UCL Eastman Dental Institute, UK	
1.1	Introduction	3
1.2	Extracellular matrices (ECMs)	5
1.3	ECM and cell interaction	9
1.4	ECM and mechanical signalling	12
1.5	Future trends	13
1.6	Conclusion	14
1.7	References	14

2	**Three-dimensional collagen biomatrix development and control**	**18**
	U. CHEEMA, UCL Institute of Orthopaedics and Musculoskeletal Science, UK	
2.1	Engineering cell-rich and matrix-rich tissues using collagen scaffolds	18
2.2	Controlling the mechanical properties of collagen	21
2.3	Architectural features: introducing elements of tissue complexity	26

v

vi   Contents

2.4	Future trends	30
2.5	References	31

3   Two- and three-dimensional tissue culture bioprocessing methods for soft tissue engineering   34
M. J. ELLIS, University of Bath, UK

3.1	Introduction	34
3.2	Bioreactor configurations	35
3.3	Selecting scaffold materials and architectures for your bioreactor	39
3.4	Mass transfer in tissue engineering bioreactors	43
3.5	Important parameters and taking measurements of bioreactor cultures	45
3.6	Tissue engineering process design	46
3.7	Future trends	48
3.8	Conclusion	49
3.9	Sources of further information and advice	49
3.10	References	50

4   Two- and three-dimensional tissue culture methods for hard tissue engineering   54
M. A. BIRCH and K. E. WRIGHT, Newcastle University, UK

4.1	Introduction	54
4.2	Culture of bone and cartilage cells	56
4.3	Cell culture parameters: bone tissue culture	60
4.4	Cell culture parameters: cartilage tissue culture	64
4.5	Two-dimensional tissue culture methods for hard tissues	67
4.6	Two-and-a-half- and three-dimensional tissue culture methods for hard tissues	68
4.7	Conclusion	71
4.8	References	71

5   Vascularisation of tissue-engineered constructs   77
B. BURANAWAT, P. KALIA and L. DI SILVIO, King's College London, Dental Institute, UK

5.1	Introduction	77
5.2	Growth of healthy vessels – embryonic vasculogenesis	79
5.3	Angiogenic diseases	81
5.4	Angiogenesis and bone formation	83

© Woodhead Publishing Limited, 2013

5.5	Cell sources for vascular tissue engineering	91
5.6	Co-culture of cells: the interactions between angiogenesis and osteogenesis	91
5.7	Strategies to induce *in vitro* prevascularisation	92
5.8	Tubular formation	94
5.9	Conclusion	94
5.10	References	96

Part II	**Standards and protocols in cell and tissue engineering**	**105**

6	**Standards in cell and tissue engineering**	**107**
	P. Tomlins, Consultant, UK	
6.1	Introduction	107
6.2	How and by whom are standards produced?	109
6.3	The importance of an agreed lexicon	109
6.4	Drivers for standardization	110
6.5	How will standards help me?	111
6.6	What standards currently exist in tissue engineering?	111
6.7	Characterization of biomaterials and biomolecules	112
6.8	Characterization of tissue scaffolds	112
6.9	Characterization of cell-seeded scaffolds	119
6.10	Manufacture, processing and storage	120
6.11	Characterization of cells and cell–surface interactions	121
6.12	Conclusion	122
6.13	References	123

7	**Principles of good laboratory practice (GLP) for *in vitro* cell culture applications**	**127**
	B. Idowu and L. Di Silvio, King's College London, Dental Institute, UK	
7.1	Introduction	127
7.2	GLP governing bodies	129
7.3	Resources required for GLP compliance	132
7.4	Characterisation	135
7.5	Standards and regulations	138
7.6	Documentation of results	141
7.7	Independent monitoring of research processes and quality assurance (QA) personnel	142
7.8	Application of GLP to human cell culture systems	143

viii   Contents

7.9	Conclusion	145
7.10	Acknowledgements	145
7.11	References	146

8   Quality control in cell and tissue engineering   **148**
I. B. WALL, University College London, UK and N. DAVIE,
University of Oxford, UK

8.1	Introduction	148
8.2	Quality control to ensure a well-defined cell therapy product	151
8.3	Commercial quality control/quality assurance in large-scale manufacture	157
8.4	Conclusion	162
8.5	References	162

9   Standardised chemical analysis and testing of biomaterials   **166**
W. CHRZANOWSKI and F. DEHGHANI, The University of
Sydney, Australia

9.1	Introduction: why we need standard methods for testing biomaterials	166
9.2	Standardised chemical analysis: when and why we assess chemistries	167
9.3	Chemical properties	168
9.4	Imaging methods for measuring porosity	173
9.5	Physical characterisation – permeability	178
9.6	Surface properties	179
9.7	Degradation and stability in physiological fluids	181
9.8	Implant–tissue interface tests	184
9.9	Limitations of current standardised testing methods	193
9.10	References	194

10   Sterilisation procedures for tissue allografts   **197**
B. J. PARSONS, Leeds Metropolitan University, UK

10.1	Introduction	197
10.2	Interaction of ionising radiation with matter	198
10.3	Sources of ionising radiation	202
10.4	Validation and international standards of sterilisation by ionising radiation	204

© Woodhead Publishing Limited, 2013

		Contents	ix
10.5	Conclusions and future trends		210
10.6	Sources of further information and advice		210
10.7	References		210

11 Commercial manufacture of cell therapies   212
I. B. WALL, University College London, UK and
D. A. BRINDLEY, University of Oxford, UK and Harvard
University, USA

11.1	Introduction: cells as therapies	212
11.2	The transition from laboratory to commercial-scale manufacture of cell therapies	220
11.3	Key regulatory requirements for commercial manufacture of cell therapies	225
11.4	Cell-based therapy versus monoclonal antibody therapies: lessons from existing biopharmaceutical manufacture	230
11.5	Conclusion	235
11.6	References	235

Index   *241*

Contributor contact details

(* = main contact)

Editor

Vehid Salih
Peninsula Dental School
    Plymouth University Schools of
    Medicine and Dentistry
The John Bull Building
Research Way
Plymouth
Devon
PL6 8BU
UK

E-mail: vehid.salih@plymouth.ac.uk

Chapter 1

Vehid Salih*
Peninsula Dental School
    Plymouth University Schools of
    Medicine and Dentistry
The John Bull Building
Research Way
Plymouth
Devon
PL6 8BU
UK

E-mail: v.salih@.ucl.ac.uk

Dominique Thomas
Department of Biomaterials &
    Tissue Engineering
UCL Eastman Dental Institute
256 Gray's Inn Road
London
WC1X 8LD
UK

Chapter 2

Umber Cheema
Institute of Orthopaedics and
    Musculoskeletal Science
UCL Division of Surgery &
    Interventional Science
Stanmore Campus
Brockley Hill
Stanmore
HA7 4LP
UK

E-mail: u.cheema@ucl.ac.uk

Chapter 3

Marianne J. Ellis
Centre for Regenerative Medicine
Department of Chemical
    Engineering
University of Bath
Claverton Down

xii Contributor contact details

Bath
BA2 7AY
UK

E-mail: m.j.ellis@bath.ac.uk

Chapter 4

Mark A. Birch* and Kathleen
Wright
Institute of Cellular Medicine
Musculoskeletal Research Group
The Medical School
Newcastle University
Newcastle upon Tyne
NE2 4HH
UK

E-mail: mark.birch@ncl.ac.uk

Chapter 5

Borvornwut Buranawat, Priya Kalia
and Lucy Di Silvio*
Biomaterials, Tissue Engineering &
Imaging Group
King's College London
Dental Institute
Guy's Hospital
London
SE1 9RT
UK

E-mail: lucy.di_silvio@kcl.ac.uk

Chapter 6

Paul Tomlins
Englefield Green
Surrey
TW20 0JY
UK

E-mail: tomlinspaul@gmail.com

Chapter 7

Bernadine Idowu and Lucy Di
Silvio*
Biomaterials, Tissue Engineering &
Imaging Group
King's College London
Dental Institute
Guy's Hospital
London
SE1 9RT
UK

E-mail: lucy.di_silvio@kcl.ac.uk;
bernadine.idowu@kcl.ac.uk

Chapter 8

Ivan B. Wall*
Advanced Centre for Biochemical
Engineering
University College London
Torrington Place
London
WC1E 7JE
UK

E-mail: i.wall@ucl.ac.uk

Natasha Davie
Centre for Accelerating Medical
Innovations
University of Oxford
Old Road Campus
Oxford
OX3 7LG
UK

E-mail: natasha@eucami.org

© Woodhead Publishing Limited, 2013

Contributor contact details   xiii

Chapter 9

Wojciech Chrzanowski
The University of Sydney
The Faculty of Pharmacy
NSW 2006
Australia

E-mail: wchrzanowski@sydney.edu.au

Fariba Dehghani
The University of Sydney
School of Chemical and
    Biomolecular Engineering
Faculty of Engineering and
    Information Technology
NSW 2006
Australia

E-mail: fariba.dehghani@sydney.edu.au

Chapter 10

B. Parsons
Faculty of Health and Social
    Sciences
Leeds Metropolitan University
Calverley Street
Leeds
LS1 3HE
UK

E-mail: b.parsons@leedsmet.ac.uk

Chapter 11

Ivan B. Wall*
Advanced Centre for Biochemical
    Engineering
University College London
Torrington Place
London
WC1E 7JE
UK

E-mail: i.wall@ucl.ac.uk

David A. Brindley
Nuffield Department of
    Orthopaedics, Rheumatology and
    Musculoskeletal Sciences
University of Oxford
Nuffield Orthopaedic Centre
Windmill Road
Oxford
OX3 7HE
UK

E-mail: david@eucami.org

© Woodhead Publishing Limited, 2013

Woodhead Publishing Series in Biomaterials

1 **Sterilisation of tissues using ionising radiations**
   Edited by J. F. Kennedy, G. O. Phillips and P. A. Williams
2 **Surfaces and interfaces for biomaterials**
   Edited by P. Vadgama
3 **Molecular interfacial phenomena of polymers and biopolymers**
   Edited by C. Chen
4 **Biomaterials, artificial organs and tissue engineering**
   Edited by L. Hench and J. Jones
5 **Medical modelling**
   R. Bibb
6 **Artificial cells, cell engineering and therapy**
   Edited by S. Prakash
7 **Biomedical polymers**
   Edited by M. Jenkins
8 **Tissue engineering using ceramics and polymers**
   Edited by A. R. Boccaccini and J. Gough
9 **Bioceramics and their clinical applications**
   Edited by T. Kokubo
10 **Dental biomaterials**
   Edited by R. V. Curtis and T. F. Watson
11 **Joint replacement technology**
   Edited by P. A. Revell
12 **Natural-based polymers for biomedical applications**
   Edited by R. L. Reiss et al
13 **Degradation rate of bioresorbable materials**
   Edited by F. J. Buchanan
14 **Orthopaedic bone cements**
   Edited by S. Deb
15 **Shape memory alloys for biomedical applications**
   Edited by T. Yoneyama and S. Miyazaki
16 **Cellular response to biomaterials**
   Edited by L. Di Silvio
17 **Biomaterials for treating skin loss**
   Edited by D. P. Orgill and C. Blanco

© Woodhead Publishing Limited, 2013

xvi Woodhead Publishing Series in Biomaterials

18 **Biomaterials and tissue engineering in urology**
Edited by J. Denstedt and A. Atala

19 **Materials science for dentistry**
B. W. Darvell

20 **Bone repair biomaterials**
Edited by J. A. Planell, S. M. Best, D. Lacroix and A. Merolli

21 **Biomedical composites**
Edited by L. Ambrosio

22 **Drug–device combination products**
Edited by A. Lewis

23 **Biomaterials and regenerative medicine in ophthalmology**
Edited by T. V. Chirila

24 **Regenerative medicine and biomaterials for the repair of connective tissues**
Edited by C. Archer and J. Ralphs

25 **Metals for biomedical devices**
Edited by M. Ninomi

26 **Biointegration of medical implant materials: science and design**
Edited by C. P. Sharma

27 **Biomaterials and devices for the circulatory system**
Edited by T. Gourlay and R. Black

28 **Surface modification of biomaterials: methods analysis and applications**
Edited by R. Williams

29 **Biomaterials for artificial organs**
Edited by M. Lysaght and T. Webster

30 **Injectable biomaterials: Science and applications**
Edited by B. Vernon

31 **Biomedical hydrogels: Biochemistry, manufacture and medical applications**
Edited by S. Rimmer

32 **Preprosthetic and maxillofacial surgery: Biomaterials, bone grafting and tissue engineering**
Edited by J. Ferri and E. Hunziker

33 **Bioactive materials in medicine: Design and applications**
Edited by X. Zhao, J. M. Courtney and H. Qian

34 **Advanced wound repair therapies**
Edited by D. Farrar

35 **Electrospinning for tissue regeneration**
Edited by L. Bosworth and S. Downes

36 **Bioactive glasses: Materials, properties and applications**
Edited by H. O. Ylänen

37 **Coatings for biomedical applications**
Edited by M. Driver

38 **Progenitor and stem cell technologies and therapies**
Edited by A. Atala

39 **Biomaterials for spinal surgery**
Edited by L. Ambrosio and E. Tanner

© Woodhead Publishing Limited, 2013

Woodhead Publishing Series in Biomaterials

40 **Minimized cardiopulmonary bypass techniques and technologies**
Edited by T. Gourlay and S. Gunaydin

41 **Wear of orthopaedic implants and artificial joints**
Edited by S. Affatato

42 **Biomaterials in plastic surgery: Breast implants**
Edited by W. Peters, H. Brandon, K. L. Jerina, C. Wolf and V. L. Young

43 **MEMS for biomedical applications**
Edited by S. Bhansali and A. Vasudev

44 **Durability and reliability of medical polymers**
Edited by M. Jenkins and A. Stamboulis

45 **Biosensors for medical applications**
Edited by S. Higson

46 **Sterilisation of biomaterials and medical devices**
Edited by S. Lerouge and A. Simmons

47 **The hip resurfacing handbook: A practical guide to the use and management
of modern hip resurfacings**
Edited by K. De Smet, P. Campbell and C. Van Der Straeten

48 **Developments in tissue engineered and regenerative medicine products**
J. Basu and J. W. Ludlow

49 **Nanomedicine: technologies and applications**
Edited by T. J. Webster

50 **Biocompatibility and performance of medical devices**
Edited by J-P. Boutrand

51 **Medical robotics: minimally invasive surgery**
Edited by P. Gomes

52 **Implantable sensor systems for medical applications**
Edited by A. Inmann and D. Hodgins

53 **Non-metallic biomaterials for tooth repair and replacement**
Edited by P. Vallittu

54 **Joining and assembly of medical materials and devices**
Edited by Y. (Norman) Zhou and M. D. Breyen

55 **Diamond-based materials for biomedical applications**
Edited by R. Narayan

56 **Nanomaterials in tissue engineering: Fabrication and applications**
Edited by A. K. Gaharwar, S. Sant, M. J. Hancock and S. A. Hacking

57 **Biomimetic biomaterials: Structure and applications**
Edited by A. Ruys

58 **Standardisation in cell and tissue engineering: Methods and protocols**
Edited by V. Salih

59 **Inhaler devices: Fundamentals, design and drug delivery**
Edited by P. Prokopovich

60 **Bio-tribocorrosion in biomaterials and medical implants**
Edited by Y. Yan

61 **Microfluidics for biomedical applications**
Edited by X.-J. James Li and Y. Zhou

© Woodhead Publishing Limited, 2013

xviii   Woodhead Publishing Series in Biomaterials

62 **Decontamination in hospitals and healthcare**
    Edited by J. T. Walker
63 **Biomedical imaging: Applications and advances**
    Edited by P. Morris
64 **Characterization of biomaterials**
    Edited by M. Jaffe, W. Hammond, P. Tolias and T. Arinzeh
65 **Biomaterials and medical tribology**
    Edited by J. P. Davim
66 **Biomaterials for cancer therapeutics: Diagnosis, prevention and therapy**
    Edited by K. Park
67 **New functional biomaterials for medicine and healthcare**
    E. P. Ivanova, K. Bazaka and R. J. Crawford
68 **Porous silicon for biomedical applications**
    Edited by H. A. Santos
69 **A practical approach to spinal trauma**
    Edited by H. N. Bajaj and S. Katoch
70 **Rapid prototyping of biomaterials: Principles and applications**
    Edited by R. Narayan
71 **Cardiac regeneration and repair Volume 1: Pathology and therapies**
    Edited by R.-K. Li and R. D. Weisel
72 **Cardiac regeneration and repair Volume 2: Biomaterials and tissue
    engineering**
    Edited by R.-K. Li and R. D. Weisel
73 **Semiconducting silicon nanowires for biomedical applications**
    Edited by J. L. Coffer
74 **Silk for biomaterials and tissue engineering applications**
    Edited by S. Kundu
75 **Novel biomaterials for bone regeneration: Novel techniques and applications**
    Edited by P. Dubruel and S. Van Vlierberghe
76 **Biomedical foams for tissue engineering applications**
    Edited by P. Netti
77 **Precious metals for biomedical applications**
    Edited by N. Baltzer and T. Copponnex
78 **Bone substitute biomaterials**
    Edited by K. Mallick
79 **Regulatory affairs for biomaterials and medical devices**
    Edited by S. Amato and R. Ezzell
80 **Joint replacement technology Second edition**
    Edited by P. A. Revell
81 **Computational modelling of biomechanics in the musculoskeletal system:
    Tissues, replacements and regeneration**
    Edited by Z. Jin
82 **Biophotonics for medical applications**
    Edited by I. Meglinski
83 **Modelling degradation of bioresorbable polymeric medical devices**
    Edited by J. Pan

© Woodhead Publishing Limited, 2013

Foreword

Progress in both the materials sciences and life sciences, especially during the past two decades, has led to a rapid expansion in regenerative medicine, of which tissue engineering is a central component. This progress has been accompanied by a paradigm change in treating tissue and organ defects from a replacement strategy towards a regenerative strategy. The latter concept relies heavily on knowledge of the biological mechanisms of regeneration, as well as on the availability of suitable support structures as vehicles for repair. The former has profited from the expansion of stem cell biology and the latter from innovative technologies in all classes of biomaterials, but particularly in polymers.

If we regard the classical tissue engineering construct as containing biomaterials, signal molecules and cells as typical components and add to this the increasingly important issue of biomechanical forces, represented, for example, by bioreactor technology, then it becomes evident that standardisation is a *conditio sine qua non* for successful clinical translation. This lies at the heart of Vehid Salih's book project, which focuses on important issues relevant for standardisation.

A corollary of the regenerative strategy is that the support structures for cell and tissue engineering should be as close to the natural 'gold standard' as possible, and this involves simulation of the extracellular matrix (ECM), which is the microenvironment of cells in the living organism. Thus, the book begins with a comprehensive view of matrix biology and presents the principal molecular players and how they are adapted to regulating cellular functions.

Even from a glance at the field of tissue engineering it is evident that standardisation is extremely difficult, as there are so many types and combinations of materials, biological signals and cells. Clarity is brought into this in the subsequent chapters, which discuss, for example, the guidance documents and directives from responsible organisations, at national, European (CEN) and international (ISO) level. Among the important topics discussed are biomaterial testing with special emphasis on physico-chemical analysis, biomaterial degradation behaviour, sterilisation and storage of biomaterials, cells, etc., and characterisation of cell–biomaterial interactions. Thus, the

xx   Foreword

reader is given a good perspective on how the various elements of tissue engineered products can be tested in such a manner as to be relevant for clinical translation. Topics, such as quality control and quality assurance in cell therapies, or the practicalities of Good Laboratory Practice (GLP) applied to cell culture are explained in a lucid way.

How the various principles of standardisation are applied to real-life cell and tissue engineering is illustrated in a number of chapters which give specific examples, such as bioreactor design for soft tissue engineering, the use of collagens, and hard tissue engineering. As the authors develop these themes, laboratory methods are given a high priority and include discussion of culture methodologies with a review of permanent cell lines, primary and progenitor cell sources. Attention is also given to some of the essential challenges in the tissue engineering field, such as promotion of vascularisation.

One of the burning questions for publisher and reviewer alike is the intended target population. To whom is the book addressed and how much background knowledge in the field is required to benefit from it? Dr Salih, who not only edits the book but is also an active scientific coauthor, has successfully brought together a group of expert authors who in their *vitae* represent the heterogeneity of the field itself. Thus, although most are colleagues from academic institutions, they cover the spectrum from pure to applied science and have invested considerable effort in demonstrating how state-of-the-art technologies in the individual sciences, whether cell culture or surface chemical analysis, must be applied in standardisation procedures. This is anything but an academic exercise, but rather involves a necessary state of mind in approaching cell and tissue engineering. The book is thus highly relevant for those, especially younger colleagues entering the field, either from the academic or industrial side. Good practical advice is given, for example, on sources of expert help, both in the form of literature citations, and reference to organisations with various types of specialisation. Moreover, this is clothed in a language which is, of course, multidisciplinary, but nevertheless not overloaded with technical jargon. Naturally, as the book was intentionally compiled in compact form, there is no claim whatsoever to be an exhaustive treatise. Thus, the reader should approach it with a view to gaining insight into how established methodologies in the materials and life sciences can be used and the resulting data interpreted in the light of standardisation criteria, this process being a pre-requisite for clinical translation. Despite the academic stimulation inevitably generated by interdisciplinary approaches, we should never lose sight of the fact that biomaterials are *per definitionem* intended for human application. The latter serves to focus our attention on the constant challenges and dangers in extrapolating from even the most sophisticated of standardisation models and technologies to the human application.

© Woodhead Publishing Limited, 2013

In summary, I consider this book to be an invaluable contribution to the topic of tissue engineering and, although compact, it will serve as a useful reference in all laboratories working in the field. In addition, it has the potential to be included in university curricula in courses on biomaterials, tissue engineering and bioengineering.

C. J. Kirkpatrick
Director, Institute for Pathology
University of Mainz
Germany

Introduction

The field of research covered by this book represents and links into the basic science platform for a wide range of current biotechnical activities commonly referred to as tissue engineering and regenerative medicine. The overall aim is to introduce concepts and current practice to a wide audience experiencing a second generation of tissue engineering innovation and promise. There has been a clear and defined hypothetical shift in regenerative medicine from using solely synthetic medical devices and tissue graft, to a more explicit approach that utilises specific biodegradable synthetic or natural scaffolds combined with cells and/or biological molecules, in order to create a functional replacement tissue in a diseased or damaged tissue site. Every era in medical research over the past 50 years involving the use of biomaterials in order to replace tissue function has been distinct and identified by particular materials and developmental successes. For example, in the 1950s, there was a predominant use of metal implants and associated devices. Throughout the 1970s and 1980s, there was a significant increase in the use of polymers and synthetic materials, and more recently, there has been a distinct and concentrated effort in the design and use of both natural and degradable scaffolds.

There has been an evolution from the use of biomaterials to simply replace non-functioning tissue to that of utilising specific materials which will nurture, in three dimensions, a fully functioning and structurally acceptable tissue. Thus, the simple need to accomplish the replacement of a functioning joint using a fully metal prosthesis during the pioneering days of Sir John Charnley in the 1960s has been markedly enhanced to concentrate on biological aspects of the damaged or diseased tissue to be replaced by repaired, or better still, totally regenerated tissue. There was a very naïve belief that materials were typically 'inert' and this is a misleading interpretation, as it became clear that materials could indeed change physically and chemically following implantation. Certainly from a biological perspective, no material should be considered inert. It has become quite apparent, therefore, that the choice of scaffold is crucial to enable the cells in question to function in an appropriate manner to produce the required extracellular matrix, and thus tissue, of a desired geometry and size and normal functional capability.

xxiv   Introduction

There are however, legislative and ethical hurdles which need to be cleared, for example, consideration and implementation of the Human Tissue Act (2004), in order to ensure the continued acceleration of this exciting field of research. Clearly, the potential health benefits to individuals and health services are vast. Stern challenges lie ahead when one considers that, for all the research groups in many different countries and the numbers of materials being investigated, very few examples of scaffolds are in clinical use at present. Some of the issues that need to be addressed include the native tissues and organs, and the variety of tissue structures. Do scientists need to consider two-dimensional culture techniques as possibly obsolete as more and more groups switch to three-dimensional bioreactor culture? Furthermore, the question of critical-sized defects comes into play, that is, do we consider microscale *vs* nanoscale; various cell types; culture systems including perfusion of scaffolds; vascularity; waste and nutrient consumption biomonitoring? These are important features of any future improvement of metrology within tissue engineering systems and the development of the next generation of scaffolds.

Researchers within the field are striving to improve understanding of cell/material interactions; the ability to control the host response; and enabling the standardisation of functional assays/protocols by limiting types of cells in use and by standardising isolation methods. This can be assessed by using the same markers of biological function and developing a national (and ultimately international) group of reference materials. The adoption of standards is vital as it will lead to innovation, reduced development, processing and manufacturing costs, and more products to market and highlight previous intellectual property problems in industry by encompassing validation and accurate measurement modalities. In due course, these considerations need to involve numerous stakeholders, namely the major research councils and government departments (DTI, DoH) as well as the main standards authorities and regulatory bodies (ASTM, ISO, BSI, NIBSC, MHRA). The long-term impact will help direct and form development of local ideas to be accepted nationally/internationally. These measures will provide 'best practice' methods in order to create safe, high-quality performance-related products. Educating regulatory authorities and end users about the technical standards can only lead to long-term benefits to the health sector and, ultimately, the end users (i.e. the patients).

It is clear that the 'promises' and objectives of tissue engineering proposed some 30 years ago have not delivered the numerous products and synthetic therapies for many of the tissue requirements of diseased or traumatised tissues. The foundation that may address this is the standardisation of the induction, development, and maintenance of differentiation/maturation of various cell types, as well as tissue matrices under *in vitro* conditions.

© Woodhead Publishing Limited, 2013

Thus, the source of cells, their ability to differentiate and proliferate, the optimisation of culture conditions, and the choice of both natural and synthetic matrices (i.e. scaffolds), on which *in vitro* as well as *in vivo* host tissues can develop, are all aspects that must be evaluated when considering standards in this field. To some extent the Food and Drug Administration (FDA) and American Society for Testing and Materials (ASTM) are making a concerted effort to establish standards and guidelines for the entire field of tissue-engineered medical products (TEMPs). This type of endeavour is slow and laboured within the EU and the rest of the world. It is true that there are available various ISO guidelines, documents and developed medical products which are scrutinised and regulated by various authorities (FDA and the Medical Devices Agency UK), but what about the standardisation of creating the TEMPs and products before they reach development and the market place? It seems that in the literature there is a lack of transparent guidelines of various methods and protocols of the developmental stages of such TEMPs.

It is my intention that this book will provide insight into some of the standard protocols and methods currently utilised in tissue engineering. I would like the book chapters to comprehensively outline the fundamentals of standard protocols in cell maintenance in culture, preservation and characterisation, as well as the technology and issues surrounding the synthesis, evaluation and characterisation of cell-based tissue-engineered products. The regulatory framework surrounding these applications for tissue-engineered products and biological applications in general would be addressed with reference to the already numerous tissue-engineered products around the world. Standardisation, regulation and validation are vital issues in all aspects of cell and tissue engineered therapies and the idea of this volume is to bring together standard protocols, issues and relevant information for these therapies with a view to bringing more products more speedily to the market. The closing chapters address both good laboratory practice (GLP) and good manufacturing practice (GMP) for cell-culture processes and encourage research laboratories worldwide to digest some of these basic principles. The work between laboratories and research groups would then be so much more comparable in the literature and relevant for stakeholders such as government agencies, regulatory bodies as well as the main consumers, the public. The book is written for a wider audience to provide awareness of the important world of standardisation in tissue engineering and provide directions for further investigation into specific topics. It will be of value to leading research groups, government agencies, regulatory bodies, and researchers and technicians at all levels across the whole range of disciplines using cell culture within the pharmaceutical, biotechnology and biomedical industries.

xxvi   Introduction

Lastly, I would like to offer my sincere gratitude to the authors of the chapters for their valuable contributions and to Woodhead Publishing for the opportunity of my first experience as an editor.

Vehid Salih
University College London

Part I
Methods for cell and tissue engineering

© Woodhead Publishing Limited, 2013

1
Fundamentals of cell and matrix biology for tissue engineering

V. SALIH, Plymouth University Schools of Medicine and Dentistry, UK (formerly at UCL Eastman Dental Institute, UK) and D. THOMAS, UCL Eastman Dental Institute, UK

DOI: 10.1533/9780857098726.1.3

Abstract: While cells are the important producers and directors of tissues, being mainly responsible for the synthesis and maintenance of the tissue, it is the extracellular matrix (ECM) that makes up a substantial part of the tissue volume and largely comprises a complex network of macromolecules. A variety of proteins, including several collagen types, are secreted locally and orchestrated into a highly organised milieu which links cells to one another as well as key ECM proteins.

Traditionally thought of as a structurally stable support material, the ECM of mammalian tissues also has a pivotal role in biological function, largely through the ability to bind multiple factors including non-collagenous proteins, growth factors, signal receptors and a large variety of adhesion molecules. It thus has a far more active role in regulating cell behaviour, influencing survival, migration, proliferation, shape, differentiation and, ultimately, cell functions. The ECM has a correspondingly complex molecular composition which will be highlighted here with relevant examples.

Key words: extracellular matrix (ECM), matrix protein, cell interaction, mechanical signalling.

1.1 Introduction

In the simplest of terms, a tissue engineering scaffold may be described as a support structure with viable cells coupled with a suitable culture environment to support the development of functional tissue. For the purposes of this chapter, a scaffold for tissue engineering purposes will be considered as a permanently placed or temporary, three-dimensional porous and permeable natural or synthetic biomaterial that is compatible with a particular mammalian cell type with respect to its bulk form as well as its degraded constituent form. Moreover, the scaffold should ideally possess appropriate mechanical properties for the developing tissue(s) in question and possess suitable physical and chemical properties to allow cell adhesion, migration, proliferation, differentiation to the cell's mature phenotype with simultaneous extracellular matrix (ECM) production and maintenance. Although extensive, these seem reasonable and sensible demands of a material when one considers the need for tissue or organ replacement.

The concept of understanding and developing improved scaffolds is inherent in the need to fully comprehend how we have moved on from the small-scale monolayer principle to the more complex three-dimensional efforts in growing cell constructs of a suitable size for clinical use. Thus, a suitable and controlled geometric cell scaffold would ultimately produce a viable tissue construct.[1]

Furthermore, the scaffold must be multifunctional. Prior to implantation, during the *in vitro* preparation phase of any cell construct, the scaffold must host the cells of interest and any extracellular matrices or additional growth/differentiation agents.[2] This is just the beginning, as mature constructs must be viable and to do this, they need to allow adequate perfusion of nutrients and waste and, if applicable, they need to allow appropriate vascularisation.[3–5] As well as the choice of adequate material – a decision often made with many assumptions and at the expense of fundamental biological understanding of the tissue matrix requirements – there is an essential need for at least an appreciation of the physiochemical properties of the material. These properties will ultimately affect the basic requirements of cell survival, growth, proliferation and matrix reorganisation, but from a functional point of view, they need to preserve appropriate gene and therefore, protein function. Among the most important proteins produced by cells are those that synthesise and maintain the highly regulatory ECM, which is vital for successful integration of any biomaterial, whether natural or synthetic and how it interplays with a maturing cell matrix during healing. On several occasions in the preceding paragraphs, the words 'matrix' or 'ECM' have been utilised and it is precisely these terms which are the focus of this chapter.

Matrix biology is the study of ECM and its communication with cells. Where adhered cells are present, matrix is also present, throughout the human body. This area of biology is gaining wide prominence as a key structural and functional regulator of cell and tissue function and can be further divided into the study of cellular microenvironments as determined by extracellular matrix (ECM) and basement membranes (BMs) in the regulation of the tissue behaviour during health and disease. This fundamental interest in matrix biology translates into several major areas, namely vascular biology and angiogenesis; the tumour microenvironment, which includes cancer progression and metastasis; genetic and acquired connective tissue diseases and, more recently, tissue engineering and stem cell biology. By studying ECM and its communication with cells we are able to determine how tissues behave in health and disease and in particular, in tissue engineering scenarios where cell–matrix interactions are paramount for the success of the newly regenerated and properly functioning tissue.

Fundamentals of cell and matrix biology for tissue engineering 5

Traditionally thought of as a structurally stable support material, the ECM of mammalian tissues also has a pivotal role in biological function, largely through its ability to bind multiple factors including non-collagenous proteins, growth factors, signal receptors and a large variety of adhesion molecules. Classically, there are four major classes of ECM macromolecules: collagens, proteoglycans, structural glycoproteins and elastin. Each group contains related proteins that are the products of unique genes and these proteins have been shown to have tissue-specific patterns of expression. The presence of a specific and a functionally important set of cell-surface receptors known as integrins are involved in direct binding to the ECM. It is also widely recognised that dysfunctional matrix components and abnormalities in ECM biosynthesis and catabolism occur in both inherited and acquired disease as well as normal wound healing. The following text highlights some important molecules and important interactions that are key players in cell–matrix and cell–scaffold exchanges.

1.2   Extracellular matrices (ECMs)

Extracellular matrices (ECMs) are secreted molecules that constitute the cell microenvironment, composed of a dynamic and complex array of glycoproteins, collagens, glycosaminoglycans and proteoglycans. ECM provides the volume, shape and structural support of many tissues *in vivo*, such as basement membrane, musculoskeletal tissues and the major organs and other tissues. *In vitro*, most animal cells can only grow when they are attached to surfaces via an ECM. ECM is also the substrate for cell migration, proliferation and differentiation. However, ECM provides much more than just mechanical and structural support, with repercussions for developmental patterning, stem cell function and differentiation as well as tumour development. ECM communicates spatial context for a variety of important signalling events by various cell-surface growth factor receptors and adhesion molecules, such as integrins. The physical properties of ECM may also have a role in the signalling process. ECM molecules can be flexible and extendable, stiff and rigid, and mechanical tension can further influence interactions with both growth factors and their receptors.[6] ECM proteins and composite structures can determine the cell behaviour, polarity, migration, differentiation, proliferation and survival by communicating with the intracellular cytoskeleton and transmission of growth factor signals. Integrins and proteoglycans are the major ECM adhesion receptors which cooperate in signalling events, determining the signalling outcomes, and ultimately the cell fate. The ECM is the product of several secreted specialised proteins of the cell or cells that comprise tissue and is composed of both structural and functional proteins arranged in a unique and tissue-specific three-dimensional structure. The ECM not only has a mechanical function

6   Standardisation in cell and tissue engineering

of accommodating its resident cells but also one of a mediator of signalling, an essential function for the proper communication not only between neighbouring cells, but between those cells and the ECM itself.

The ECM provides the main basis of structural support to the resident cells; its existence enables them to form a network of connections with the ECM and adjacent cells. In order for cells to grow, mechanisms such as migration and adhesion are two important factors which need to occur and are provided by the ECM encouraging cell attachment and acting as the substrate for cell migration. Moreover proteins within the ECM can determine the fate of the cells by influencing cell differentiation and cell viability via interaction with growth factors and the intracellular proteins, in particular the actin cytoskeletal filaments. The ECM is thus a highly specialised construct and manipulates its own microenvironment. An example of a major ECM protein that has the ability to bind to multiple proteins influencing a large array of biological functions is fibronectin which is produced by fibroblasts and is capable of binding to collagens and heparin sulphate proteoglycans.[7]

1.2.1   Collagen

Collagen is the major fibrous protein in the human body and is the most abundant protein in nature. Although it has been widely assumed that the fossilisation process results in the destruction of all organic material in less than one million years,[8] researchers have successfully extracted and sequenced collagen protein from a *Tyrannosaurus rex* fossil recovered from eastern Montana, USA.[9] Indeed, the genetic structure of collagen protein is highly conserved in modern animals, and it is durable and resistant to long-term degradation.[10]

Collagen contributes about 30% of the total human body protein. This fibrous protein connects and supports tissues including skin, bone, muscles, tendons, cartilage and all organs. Collagen is the main component of the ECM. The basic structure of collagen consists of three polypeptide chains forming a triple helix which are then ordered to form fibrils. There are almost 30 genetically unique types of collagen,[11] all of which exhibit specific properties suited to their physiological purpose. Collagen types I, III, V and XI have the ability to self-assemble into fibrils, while others form in the extracellular environment via pro-collagen metalloproteinases and the cleavage of terminal peptides.[12-14] The main function of type I collagen found within tissues such as tendons, bone and ligaments is to maintain the integrity of the tissue structure; this is displayed in its ability to withstand cyclic mechanical loading resisting various mechanical loads, including compressive forces, shear and tensile forces.

© Woodhead Publishing Limited, 2013

A common route for sourcing collagen is via animal tissues from bovine, porcine and murine sources. The most common tissues used are skin and tendon, for example, type I bovine collagen for medical applications is used from the Achilles tendon. Although they are the most common sources, the use of animal tissues raises concerns regarding the ability to completely purify collagens to avoid transmission of disease, thus emphasising the argument for the use of recombinant sources of human collagen.[15] The ECM mainly consists of type I collagen representing approximately 90% of the ECM's dry weight within many tissue types;[16] however other types are found in smaller quantities in tissues, collagen type IV for example. This is present in the basement membrane of vascular tissues and provides a unique structural component with properties that have been found to drive cellular adhesion and interaction. Further, collagen type IV has a proven high affinity for endothelial cells, highlighting the beneficial medical application seen in the coating of instruments and scaffolds intended for vascular contact in order to improve the biocompatibility of the devices.[17–19]

The collagen type found within various tissues is highly dependent on the physiological factors required for the tissue or organ to perform its function. Thus, collagen type III is found within the urinary bladder where the main property required is increased flexibility, in contrast to other structures requiring greater strength and stiffness, as seen in collagen type I in bone and tendons. Collagen type VI provides a crucial interaction component between larger structural proteins such as collagen type I and glycosaminoglycans forming a gel-like phase within the ECM.[19]

The fact that collagen is a naturally derived biomaterial rarely causing an acute inflammatory response makes it a promising biomaterial for use in tissue engineering with the aim of developing novel scaffold constructs. The use of collagen also provides other opportunities for the production of composites with the addition of glycosaminoglycans, ceramics and other synthetic biodegradable polymers.[20]

1.2.2 Fibronectin

In terms of quantity, fibronectin (a dimeric molecule) is the second most abundant molecule located throughout the ECM and is one of the most studied and therefore understood molecules of mammalian ECM.

Fibronectin is a significant component found in the ECM of embryos and is crucial to the successful development of vascular structures. In particular, fibronectin possesses ligands that have displayed the ability to bind to many cell types[21–24] and allows for advantageous uses within cell-culture techniques where coated fibronectin flasks are used to promote adhesion for specific cell types, or in tissue engineering methods where fibronectin-coated synthetic scaffolds are used to induce scaffold integration and biocompatibility.

© Woodhead Publishing Limited, 2013

8   Standardisation in cell and tissue engineering

An important element of fibronectin is the arginine-glycine-aspartate motif (RGD): a tripeptide fundamental sequence in cell adhesion through binding integrin $\alpha5\beta1$, a key integrin for fibronectin matrix construction.[19] Fibronectin can form both linear fibrils and branched networks among surrounding cells and the production of such networks is continuous where fibrils have been found to become detached when an applied shear force occurs.[25] Continuous polymerisation of fibronectin is imperative in holding matrix integrity at cell surfaces,[26] and this occurs essentially through insoluble and soluble phases where fibronectin fibrils are transformed from the insoluble deoxycholate detergent (DOC) solubility form to DOC insolubility detergent form in order to provide a structurally stable matrix.[26] The localised high concentration of receptors bound to the dimeric units of fibronectin has been shown to encourage interactions between multiple fibronectin units. The initial assembly of the matrix involves binding of integrin receptors to the fibronectin dimers; further binding of multiple fibronectin molecules occur via the N-terminal domains.[7]

1.2.3 Laminin

Another important ECM molecule which has profound functional roles, and which is mainly present in the basement membranes of tissues, as well as being partly responsible for providing the tensile strength of the tissue, is laminin. Laminin consists of three subunits – α, β and γ – which come together to form a characteristic cross-pattern that can bind to other laminins as well as proteoglycans and other ECM proteins.[27] For example, vitronectin can bind to and regulate components of the plasminogen activator signal complex, in addition to its cell adhesion duties.[28]

Laminin is a trimeric polypeptide prominently located in basement membrane ECMs and is known for its key involvement in the early developmental phase[29] influencing cell migration and differentiation and its ability to facilitate self-assembly. Laminin is a complex adhesion protein that presents many forms as a result of the particular peptide chain orientation containing α, β and γ chains.[30] To date 5α, 3β and 3γ chains have been identified forming 19 unique isoforms with each laminin isoform expressing specific signals. Laminin 111 is composed of 1α, 1β and 1γ chains expressed within mammary glands; it initiates cell differentiation[31] and also supports the growth of neurites in human mesenchymal stem cells.[32] Laminin 411, however, facilitates the adhesion of platelet and vascular endothelial cells,[33] which highlights the importance of identifying the specific function of each laminin isoform.

Laminin presents a high binding affinity for the likes of type IV collagen, heparin and cell surfaces. Laminin plays a fundamental role in maintaining vascular structures[29]; for this reason its presence is particularly applicable to tissue engineering scaffold devices in terms of scaffold production where

© Woodhead Publishing Limited, 2013

Fundamentals of cell and matrix biology for tissue engineering 9

these features are not innate to the nominated scaffold material. Laminin has been associated with growth factor mimicking sequences, and upon degradation these sequences, which are similar to the epidermal growth factor and are capable of influencing cell differentiation and proliferation, become available to surrounding cells.[34]

The laminins can self-assemble, bind to other matrix macromolecules, and have unique and shared cell interactions mediated by integrins, dystroglycan and other receptors. Through these interactions, laminins contribute dynamically to cell differentiation, cell shape and movement, maintenance of tissue phenotypes and promotion of tissue survival.[35]

1.2.4 Glycosaminoglycans

Glycosaminoglycans (GAGs) play a vital role in binding to cytokines and growth factors required for cell growth as well as being able to bind to a vast number of cell-surface receptors. The presence and mixture of GAGs in the ECM is dependent on the tissue location. GAGs that are heparin-rich provide numerous interactions with cell-surface receptors and growth factors. This highlights the beneficial application of glycosaminoglycans within the ECM, and their ability to retain water elicits the gel-like property within the ECM environment. There are a number of GAGs, including heparin sulphate and chondroitin sulphates A and B. Other dominant GAGs include aggrecan, found extensively in articular (hyaline) cartilage, and hyaluronic acid, which is at its highest concentration in fetal tissues and is most commonly used *in vitro* as a coating/additive to biomaterials when its ability to initiate cell growth and repair through scaffold tissue engineering approaches is being investigated.

Heparan sulphate proteoglycans (HSPGs) are proteoglycans found in ECM with multiple heparin sulphate (HS) side chains covalently coupled to the core protein. HSPGs present in the matrix include such molecules as perlecan, agrin, collagen type XI, syndecans and glypicans.[36] The three former ECM proteins are actively secreted into the ECM, while the latter (syndecans and glypicans) are cleaved from the cell surface by proteases and phospholipases, respectively.[37,38] Secreted HSPGs bind to most structural proteins in the ECM via a combination of protein–protein interactions as well as HS–protein interactions.

1.3 ECM and cell interaction

1.3.1 ECM and integrins

Much cell behaviour is regulated by adhesion to an ECM, including cell proliferation, differentiation, apoptosis, and importantly, gene expression, which leads to protein assembly (or disassembly) and function.

© Woodhead Publishing Limited, 2013

Cell adhesion involves binding and clustering of integrin receptors to immobilised ECM, and active spreading of the cells against the substrate. Integrin signalling, cell shape and cytoskeletal structure coordinate to control endothelial cell proliferation, differentiation and apoptosis.[39–42] These findings provide a foundation for recognising that cell adhesion is not a simple signalling event determined by binding of integrin to its ligand, but instead a complex interplay between the biochemical signals of integrins and structural changes associated with cell spreading. Cells respond to the mechanical and biochemical changes in ECM through the crosstalk between integrins and the actin cytoskeleton. Integrins are heterodimeric transmembrane receptors composed of eighteen α subunits and eight β subunits that can be non-covalently assembled into 24 combinations. The integrin dimers bind to an array of different ECM molecules with overlapping binding affinities, as summarised in a review by Alam *et al.*[43] Therefore, the specific integrin expression patterns by a cell dictate which ECM substrate the cell can bind[44] and the composition of integrin adhesomes determines the downstream signalling events, thus the eventual cell behaviour and fate. Integrins have a unique ability to respond to the molecular composition and physical properties of the ECM and integrate both mechanical and chemical signals through direct association with the cytoskeleton, which also determines the selection of specific integrin species to be involved. Integrin recognises and binds to the RGD motif, which was first discovered in fibronectin (FN) but later found in many other ECM proteins including laminin, tenascin, vitronectin and thrombospondin, to name a few.[45] The evolutionarily conserved three residue motifs, Arg-Gly-Asp, efficiently serve as the attachment site for integrin-mediated cell adhesion. Both the α and β subunits of integrins bind to RGD sequences and the specificity of integrin binding to different matrix proteins is determined, in part, by other amino acids surrounding the RGD sequence.[45] It has been shown that a density of RGD motif sufficiently high to allow a precise spatial distribution pattern of integrins seems to be required to initiate an optimal cellular response,[39] and spacing between adhesive ligand molecules in a 10–200 nm range seems to mimic physiological properties at focal adhesions.[46] Not surprisingly, the short synthetic peptide containing the RGD sequences has been explored to regulate integrin-mediated cell migration, growth, differentiation and apoptosis, as a new therapeutic agent for thrombosis, osteoporosis and cancer.[45] Numerous groups are now investigating how these changes in cell shape exert their effects at the molecular and transcriptional level, and continuing to explore numerous technological opportunities to address other questions about how ECM structure can regulate cell function.

1.3.2 ECM and growth factor signalling

Firstly, ECM serves as a local storage reservoir for growth factors. Many proteins within the ECM have dual binding sites: (i) for cell adhesion, and (ii) for growth factors, allowing local concentration of the growth factors near to their cell-surface receptors and cell adhesion sites, thus forming a higher concentration of their signal. Examples of such growth factors are the vascular endothelial growth factor (VEGF) and fibroblast growth factor (FGF), both of which bind to HSPGs; once cleaved by the heparin enzyme, the result produces soluble ligands.[47] ECM proteins play crucial and complex roles during cell-surface receptor signalling and patterning.[48,49] Both FGFs and VEGFs bind to HSPG and can be cleaved from the glycosaminoglycan components of HSPG by heparanase and released as soluble ligands. However, ECM-bound growth factors do not have to be released in soluble form to function. In fact, it is well established that FGFs actually bind to their receptors with HS as a cofactor, with the HSPG 'presenting' these ligands during signalling. Other examples are FN and vitronectin, both of which bind to hepatocyte growth factor (HGF) and form complexes with the HGF receptor and integrins.[36] In this regard, ECM can also function as an organising centre of the signalling complex on the cell surface.

The binding of ECM with growth factors typically involves specific domains of ECM proteins and results in the significant modulation of signalling activities. For example, *Drosophila* collagen IV binds to Dpp, a bone morphogenetic protein (BMP) homologue, and enhances interactions with BMP receptors.[50] Collagen type II, the dominant collagen from articular cartilage, contains a domain which binds to transforming growth factor $\beta 1$ (TGF-$\beta 1$) and BMP-2, acting as a negative regulator for these essential chondrogenic growth factors.[51]

Many ECM proteins, including laminin, tenascin, thrombospondin and fibrillin, contain epidermal growth factor (EGF)-like domains which can directly bind to an EGF receptor as soluble ligands and modulate its signalling.[52] FnIII domains of NCAM bind directly to FGFR1 and can induce ligand-independent receptor phosphorylation.[53] However, further investigation is needed to determine if there is any difference between the ECM-associated growth factor-like ligands (whether acting from the anchored solid-matrix or as a released soluble form) and the canonical ligands in terms of the specific signalling outcome.

1.3.3 ECM and proteases

A fundamental element of tissue repair and remodelling mechanisms *in vivo* is the degradation of the ECM allowing for the movement of cells, the end process being the construction of a newly formed matrix. This process

12   Standardisation in cell and tissue engineering

is regulated via proteases and heparanase. Examples of two proteases involved in the remodelling of the ECM are the matrix metalloproteinase family (MMP) and a disintegrin and metalloproteinase with thrombospondin motif (ADAMTS).[54]

There are 24 MMP proteins, six of which, however, are associated with the cell membrane via anchors, and it is these six proteins that play a vital role in the degradation of the ECM, with the additional ability to cleave other proteins such as growth factors.[55] MMPs are produced in an inactive form (in zymogens) which requires transformation into its active form. As well as the presence of MMP inhibitors, this process in turn helps to regulate the rate of ECM degradation as an uncontrolled increase in activity would result in compromised tissue integrity.[56]

There are 19 ADAMTS proteins, which exhibit specified targets in the processing of the ECM compared to MMPs. For example, ADAMTS2 can specifically cleave pro-collagen into collagen, encouraging the formation of fibrils.[57]

1.4   ECM and mechanical signalling

The biology of cell–matrix interactions is directly relevant to tissue and organ engineering. Naturally occurring tissues have inspired artificial matrices, or biomimetics, to strive to achieve the scale, composition and material properties of ECM to guide cell differentiation, migration and survival.[58] Developing nanoscale scaffolds is also emerging as a state-of-the-art technology with several ideas utilising the natural uptake of substances from the ECM into cells without the need for active uptake. This passive transfer of important molecules to cells via the ECM could revolutionise cell engineering and the manipulation of function, as the cytoskeleton may be redundant in such cases. A modified electrospinning technique has resulted in a porous nanofibrous biomaterial conducive to cell invasion and nutrient diffusion; when modified with chondroitin sulphate, this scaffold was optimised for cartilage formation.[59] In a study on skin epidermal stem cells,[60] the authors found that on collagen I-coated, acrylamide-based gels, the differentiation of the cells corresponded to the stiffness of the material. However, and perhaps unexpectedly, on polydimethylsiloxane matrices of the same stiffness range, the proportion of cells expressing a keratinocyte marker did not change with stiffness. The different responses to the surfaces were attributed to differences in the porosity of the materials. Thus the cells responded to the mechanical cues they derived from the ECM rather than the ECM stiffness itself.

Analysis of the integrin repertoire of human mesenchymal stem cells and its modulation during stem cell differentiation on matrices of varying stiffness showed that osteogenic or adipogenic differentiation typically altered the integrin pattern.[61] In tumours, the stiffness conveyed to cells by ECM has been found to be crucial in regulating cell behaviour. In a creative approach to the study of cell behaviour on so-called micro-polyacrylamide

© Woodhead Publishing Limited, 2013

Fundamentals of cell and matrix biology for tissue engineering 13

channels, matrix stiffness and pore size (confinement) were independently controlled.[62] This study concluded that cells confined to narrow channels, as probably found in dense fibrillar ECM, migrated faster compared with those cells confined to wider channels, which showed a steady increase in migration speed with stiffness. This behaviour was linked to non-muscle myosin II and the potential polarisation of traction forces in narrow pores. Such external influences seem to have a long-term effect on cell behaviour, probably through the induction of conserved cellular pathways. Culture of lung fibroblasts on stiff substrates revealed their initial cell plasticity through conversion to myofibroblasts, but also a phenomenon termed mechanical memory,[63] that is, myofibroblast phenotype, was retained for several subsequent passages on soft substrates.

1.5 Future trends

Interactions between tissue-engineered scaffolds, cells and their natural ECM affect the cellular phenotype as well as the secretion of important soluble factors that provide local cues which would lead to successful neo-tissue being regenerated.[64] Evaluating these complex interactions has been difficult. One must note that the communication between matrices and cells is a two-way process, with research effort focussed on the effect of matrices and ECM on cells, but relatively less effort centred on the effects of cells and their physiological processes on the matrices and/or scaffolds. Despite the recent increase in available information, many questions still remain, and continued research is required to build on and refine our current knowledge of the ECM, utilising in particular the rapidly improving new technologies in microscopy and imaging. The molecular nature of ECM-induced receptor complexes in the membrane and their movement upon activation can be monitored by methods such as single-molecule tracking using nanoparticles, such as nano-dots; fluorescence energy transfer technique; or high-resolution live cell imaging. These studies will reveal more potential therapeutic opportunities in ECM remodelling. The requirement might be finding specific targets from the unique microenvironment of the diseased cells or tissues.

ECM structural defects exhibit specific complications making such defects difficult to treat (damage to articular cartilage in particular is a good example of this). Although the regenerative capacity of stem cells is being explored in many different tissues, it is clear that the injured microenvironment loses consistency with scarring, inhibiting a truly regenerative scenario and reverting to a reparative situation. In conclusion, ECM components are vital proteins as well as fundamental therapeutic targets to prevent abnormal growth factor activities and have an interventional role in multiple signalling pathways. Further research and, indeed, investment is required to aid the treatment and management of trauma and degenerative diseases within tissue engineering applications.

© Woodhead Publishing Limited, 2013

1.6 Conclusion

This rather brief look at the breadth and complexity of ECM molecules and some related research highlights the functional continuity of ECM with the cell and the considerable crosstalk that results between them. As well as providing structural integrity and cell shape, the ECM offers a distinct role in growth factor regulation in terms of sequestration as well as activation of important receptors. Furthermore, mechanosensing, anchorage, spreading, migration and proliferation of stem cell niches are all paramount functions of the ECM. Together, these events have profound effects on the protein turnover within the cell. The cell–matrix interface, in particular, provides a crucial signalling link that regulates most aspects of cell behaviour, from development to differentiation and apoptosis. Whereas the traditional emphasis of the ECM field on its structural relevance has by no means diminished, there is now a strong focus in several new areas, including biomaterial development and tissue engineering. Collectively, this research offers a platform on which to build a strong and translational network, such that researchers are able to modify both mechanical and chemical signals to cells, and establish cohesive and focussed bioengineering of artificial tissues and organs using mechanical and adhesive cues.

1.7 References

1. Brown RA and Phillips JB (2007) Cell responses to biomimetic protein scaffolds used in tissue repair and engineering. *Int Rev Cytol* **262**: 75–150.
2. Agrawal CM and Ray RB (2001) Biodegradable polymeric scaffolds for musculoskeletal tissue engineering. *J Biomed Mater Res* **55**: 141–150.
3. Langer R and Vacanti JP (1993) Tissue engineering. *Science* **260**(5110): 920–926.
4. Ko HC, Milthorpe BK and McFarland CD (2007) Engineering thick tissues – the vascularisation problem. *Eur Cell Mater* **14**: 1–18.
5. Landman KA and Cai AQ (2007) Cell proliferation and oxygen diffusion in a vascularising scaffold. *Bull Math Biol* **69**(7): 2405–2428.
6. Cox TR and Erler JT (2011) Remodeling and homeostasis of the extracellular matrix: implications for fibrotic diseases and cancer. *Dis Model Mech* **4**(2): 165–178.
7. Ruoslahti E (1988) Fribronectin and its receptors. *Ann Rev Biochem* **57**: 375–413.
8. Lindahl T (1993) Recovery of antediluvian DNA. *Nature* **365**: 700.
9. Schweitzer MH, Suo Z, Avci R, Asara JM, Allen MA, Arce FT and Horner JR (2007) Analyses of soft tissue from Tyrannosaurus rex suggest the presence of protein. *Science* **316**: 277–280.
10. Tuross N and Stathoplos L (1993) Ancient proteins in fossil bones. *Methods Enzymol.* **224**: 121–129.
11. Humes DJ (2002) Building collagen molecules, fibrils, and suprafibrillar structures. *J Struct Biol* **137**: 2–10.

Fundamentals of cell and matrix biology for tissue engineering   15

12. Fratzl P (2003). Cellulose and collagen: from fibres to tissues. *Curr Opin Colloid In* **8**: 32–39
13. Pachence JM (1996) Collagen-based devices for soft tissue repair. *J Biomed Mater Res* **33**: 35–40.
14. Kadler KE, Holmes DF, Trotter JA and Chapman JA (1996) Collagen fibril formation. *Biochem J* **316**: 1–11.
15. Yang C, Hillas PJ, Báez JA, Nokelainen M, Balan JA, Tang J, Spiro R and Polarek JW (2004) The application of recombinant human collagen in tissue engineering. *BioDrugs* **18**: 103–119.
16. Roberts S, Menage J, Sandell LJ, Evans EH and Richardson JB (2009) Immunohistochemical study of collagen types I and II and procollagen IIA in human cartilage repair tissue following autologous chondrocyte implantation. *Knee* **16**: 398–404.
17. Bernard MP, Myers JC, Chu ML and Prockop DJ. (1983) Structure of a cDNA for the pro alpha 2 chain of human type I procollagen. Comparison with chick cDNA for pro alpha 2(I) identities structurally conserved features of the protein and the gene. *Biochemistry* **22**: 1139–1145.
18. Piez KA (1984). *Molecular and Aggregate Structures of the Collagens*. New York: Elsevier.
19. Yurchenco P, Birk DE and Mecham RP (1994) *Extracellular Matrix Assembly and Structure*. New York: Academic Press.
20. Pachence JM (1996) Collagen-based devices for soft tissue repair. *J Biomed Mater Res* **33**: 35–40.
21. McPherson T and Badylak SF (1998) Characterisation of fibronectin derived from porcine small intestinal submucosa. *Tiss Eng* **4**: 75–83.
22. Miyamato T, Takahashi S, Ito H, Inagaki H and Noishiki Y (1989) Tissue biocompatibility of cellulose and its derivatives. *J Biomed Mater Res* **23**: 125–133.
23. Schwarzbauer JE (1999) Basement membranes: putting up the barriers. *Curr Biol* **9**: R242–R244.
24. Schwarzbauer JE (1991) Fibronectin: from gene to protein. *Curr Opin Cell Biol* **3**(5): 786–791.
25. Engler AJ, Chan M, Boettiger D and Schwarzbauer JE. 2009. A novel mode of cell detachment from fibrillar fibronectin matrix under shear. *J Cell Sci* **122**: 1647–1653.
26. Sottile J and Hocking DC (2002) Fibronectin polymerization regulates the composition and stability of extracellular matrix fibrils and cell-matrix adhesions. *Mol Biol Cell* **13**: 3546–3559.
27. Colognato H and Yurchenco PD (2000) Form and function: the laminin family of heterotrimers. *Dev Dyn* **218**: 213–234.
28. Preissner KT and Seiffert D (1998) Role of vitronectin and its receptors in haemostasis and vascular remodelling. *Thromb Res* **89**: 1–21.
29. Li S, Harrison D, Carbonetto S, Fassler R, Smyth N, Edgar D and Yurchenco PD (2002) Matrix assembly, regulation and survival functions of laminin and its receptors in embryonic stem cell differentiation. *J Cell Biol* **157**: 1279–1290.
30. Timpl R and Brown JC (1996) Supramolecular assembly of basement membranes. *Bioessays* **18**(2): 123–132.
31. Hallman R, Horn N, Selg M, Wendler O, Pausch F and Sorokin LM (2005) Expression and function of laminins in the embryonic and mature vasculature. *Physiol Rev* **85**: 979–1000.

16 Standardisation in cell and tissue engineering

32. Mruthyunjaya S, Manchanda R, Godbole R, Pujari R, Shiras A and Shastry P (2010) Laminin-1 induces neurite outgrowth in human mesenchymal stem cells in serum/differentiation factors-free conditions through activation of FAK–MEK/ERK signaling pathways. *Biochem Biophys Res Commun* **391**: 43–48.

33. Fujiwara H, Kikkawa Y, Sanzen N and Sekiguchi K (2001) Purification and characterization of human laminin-8. Laminin-8 stimulates cell adhesion and migration through α3β1 and α6β1 integrins. *J Biol Chem* **276**: 17550–17558.

34. Schenk S, Hintermann E, Bilban M, Koshikawa N, Hojilla C, Khokha R and Quaranta V (2003) Binding to EGF receptor of a laminin-5 EGF-like fragment liberated during MMP-dependent mammary gland involution. *J Cell Biol* **161**: 197–209.

35. Colognato H and Yurchenco PD (2000) Form and function: the laminin family of heterotrimers. *Dev Dyn* **218**: 213–234.

36. Kim S-H, Turnbull J and Guimond S (2011) Extracellular matrix and cell signalling: the dynamic cooperation of integrin, proteoglycan and growth factor receptor. *J Endocrin* **209**: 139–151.

37. Brunner G, Metz CN, Nguyen H, Gabrilove J, Patel SR, Davitz MA, Rifkin DB and Wilson EI (1994) An endogenous glycosylphosphatidylinositol-specific phospholipase D releases basic FGR-Heparin Sulphate proteoglycan complexes from human bone marrow cultures. *Blood* **83**: 2115–2125.

38. Manon-Jensen T, Itoh Y and Couchman JR (2010) Proteoglycans in health and disease: the multiple roles of syndecan shedding. *FEBS J* **277**: 3876–3889.

39. Chen CS, Mrksich M, Huang S, Whitesides GM and Ingber DE (1997) Geometric control of cell life and death. *Science* **276**: 1425–1428.

40. Roberts CS, Chen CS, Mrksich M, Ingber DE and Whitesides GM (1998) Using self-assembled monolayers presenting GRGD and EG3OH groups to characterize long-term attachment of bovine capillary endothelial cells to surfaces. *J Am Chem Soc* **120**: 6548–6555.

41. Dike LE, Chen CS, Mrksich M, Tien J, Whitesides GM and Ingber DE (1999) Geometric control of switching between growth, apoptosis, and differentiation during angiogenesis using micropatterned substrates. *In Vitro Cell Dev Biol Anim* **35**: 441–448.

42. Chen CS, Alonso JL, Ostuni E, Whitesides GM and Ingber DE (2003) Cell shape provides global control of focal adhesion assembly. *Biochem Biophys Res Comm* **307**: 355–361.

43. Alam N, Goel HL, Zarif MJ, Butterfield JE, Perkins HM, Sansoucy BG, Sawyer TK and Languino LR (2007) The intergrin-growth factor receptor duet. *J Cellular Physiol* **213**: 649–653.

44. Hemler ME and Lobb RR (1995) The leukocyte beta 1 integrins. *Curr Opin Hematol* **2**: 61–67.

45. Ruoslahti E (1996) RGD and other recognition sequences for integrins. *Ann Rev Cell Dev Biol* **12**: 697–715.

46. Jiang F, Horber H, Howard J and Muller DJ (2004) Assembly of collagen into microribbons: effects of pH and electrolytes. *J Struct Biol* **148**: 268–278.

47. Patel VN, Knox SM, Likar KM, Lathrop CA, Hossain R, Eftekhari S, Whitelock JM, Elkin M, Vlodavsky I and Hoffman MP (2007) Heparanase cleavage of perlecan heparan sulfate modulates FGF10 activity during ex vivo submandibular gland branching morphogenesis. *Development* **134**: 4177–4186.

© Woodhead Publishing Limited, 2013

Fundamentals of cell and matrix biology for tissue engineering 17

48. Kirkpatrick CA, Dimitroff BD, Rawson JM and Selleck SB (2004) Spatial regulation of Wingless morphogen distribution and signalling by Dally-like protein. *Developmental Cell* **7**: 513–523.

49. Kreuger J, Perez L, Giraldez AJ and Cohen SM (2004) Opposing activities of Dally-like glypican at high and low levels of Wingless morphogen activity. *Dev Cell* **7**: 503–512.

50. Wang X, Harris RE, Bayston LJ and Ashe HL (2008) Type IV collagens regulate BMP signalling in Drosophila. *Nature* **455**: 72–77.

51. Garcia AJ, Coffinier C, Larrain J, Oelgeschlager M and De Robertis EM (2002) Chordin-like CR domains and the regulation of evolutionarily conserved extracellular signaling systems. *Gene* **287**: 39–47.

52. Schenk S, Hintermann E, Bilban M, Koshikawa N, Hojilla C, Khokha R and Quaranta V (2003) Binding to EGF receptor of a laminin-5 EGF-like fragment liberated during MMP-dependent mammary gland involution. *J Cell Biol* **161**: 197–209.

53. Kiselyov VV, Skladchikova G, Hinsby AM, Jensen PH, Kulahin N, Soroka V, Pedersen N, Tsetlin V, Poulsen FM, Berezin V and Bock E (2003) Structural basis for a direct interaction between FGFR1 and NCAM and evidence for a regulatory role of ATP. *Structure* **11**: 691–701.

54. Yong VW (2005) Metalloproteinases: mediators of pathology and regeneration in the CNS. *Nat Rev Neurosci* **6**: 931–944.

55. Mott JD and Werb Z (2004) Regulation of matrix biology by matrix metalloproteinases. *Curr Opin Cell Biol* **16**: 558–564.

56. Overall CM and Lopez-Otin C (2002) Strategies for MMP inhibition in cancer: innovations for the post-trial era. *Nat Rev Cancer* **2**: 657–672.

57. Colige A, Sieron AL, Li SW, Schwarze U, Petty E, Wertelecki W, Wilcox W, Krakow D, Cohn DH, Reardon W, Byers PH, Lapière CM, Prockop DJ and Nusgens BV (1999) Human Ehlers–Danlos syndrome type VII C and bovine dermatosparaxis are caused by mutations in the procollagen I N-proteinase gene. *Amer J Hum Genet* **65**: 308–317.

58. Kim DH, Provenzano PP, Smith CL and Levchenko A (2012) Matrix nanotopography as a regulator of cell function. *J Cell Biol* **197**: 351–360.

59. Coburn JM, Gibson M, Monagle S, Patterson Z and Elisseeff JH (2012) Bioinspired nanofibers support chondrogenesis for articular cartilage repair. *Proc Natl Acad Sci* **109**: 10012–10017.

60. Trappmann B, Gautrot JE, Connelly JT, Strange DGT, Li Y, Oyen ML, Cohen Stuart MA, Boehm H, Li B, Vogel V, Spatz JP, Watt FM and Huck WTS (2012) Extracellular-matrix tethering regulates stem-cell fate. *Nat Mater* **11**: 642–649.

61. Frith JE, Mills RJ, Hudson JE and Cooper-White JJ (2012) Tailored integrin-extracellular matrix interactions to direct human mesenchymal stem cell differentiation. *Stem Cells Dev* **21**: 2442–2456.

62. Pathak A and Kumar S (2012) Independent regulation of tumor cell migration by matrix stiffness and confinement. *Proc Natl Acad Sci.* **109**: 10334–10339.

63. Balestrini JL, Chaudhry S, Sarrazy V, Koehler A and Hinz B (2012) The mechanical memory of lung myofibroblasts. *Integr Biol (Camb)* **4**: 410–421.

64. Lutolf MP and Hubbell JA (2005) Synthetic biomaterials as instructive extracellular microenvironments for morphogenesis in tissue engineering. *Nat Biotechnol* **23**: 47–55.

2
Three-dimensional collagen biomatrix development and control

U. CHEEMA, UCL Institute of Orthopaedics and
Musculoskeletal Science, UK

DOI: 10.1533/9780857098726.1.18

Abstract: Collagen is the predominant extracellular matrix protein in the mammalian body. In particular, collagen plays a critical role in providing matrix-rich tissues, predominantly the musculo-skeletal system, with appropriate mechanical properties. The density, orientation, packing and fibril diameter of collagen is well regulated in the body. This affects the mechanical properties, diffusion properties, presence of physical stiffness gradients and chemical gradients, which all exist within tissues. Introducing these parameters into tissue-engineered collagen scaffolds is critical if we are to engineer biomimetic tissues. In this chapter, we will overview current engineering strategies to control some of these parameters with reference to how tissue function is affected.

Key words: collagen, three-dimensional engineering, plastic compression.

2.1 Engineering cell-rich and matrix-rich tissues using collagen scaffolds

Collagen is the predominant extra-cellular matrix protein component of most connective tissues within the mammalian body, comprising one-third of all protein found within tissues. Due to the structural role it plays in tissues, it is highly conserved among species and has low immunogenicity, making it an ideal scaffold material to study *in vitro* cell behaviour in three dimensions, as well as providing a three-dimensional (3D) scaffold for tissue-engineering applications.

There are currently 29 known types of collagen, but type I collagen is the most prominent type in the mammalian body, and is widely used as an *in vitro* 3D scaffold. The main mechanism used to obtain type I collagen is extraction from native tissues using acetic acid. Extraction encompasses the entire range from de-cellularisation of collagenous tissues preserving the native architecture to the complete breakdown into collagen molecules which can later be reconstituted into their native fibrillar structure.

Three-dimensional collagen biomatrix development and control 19

Collagens are extracellular and have a mainly structural role. Critical parameters including density, packing and orientation result in distinctively varying mechanical properties in tissues such as bone, skin, tendon and cartilage. In terms of representative biomatrix development, to mimic the native 3D tissue environment, these parameters need to be controlled and engineered *in vitro*.

Collagen type I scaffolds are an attractive 3D scaffold for *in vitro* studies, and cell–matrix interactions closely mimic those found *in vivo*. Most of the work on collagen scaffolds that has been done to date uses the standard hydrogel method, in which cells are suspended within low density collagen hydrogels and their behaviour observed. This method is extremely useful, but to mimic tissues more realistically, layers of complexity are being added in many approaches, including controlling specific elements of collagen fibril architecture, including packing and orientation; building gradients of collagen; and the addition of specific architectural features including topographical features and addition of extra proteins.

Other natural hydrogel materials, such as alginate hydrogels, are used extensively in tissue engineering applications. A major drawback to the use of such natural, but non-mammalian based materials, however, is the cell's inability to interact directly with the matrix. Some methods introduce or tag RGD-containing cell adhesion ligands to allow cell interaction with the alginate matrix (Rowley *et al.*, 1999). Using a native, well-conserved mammalian protein like collagen allows native cell–matrix interactions within a tissue-engineered scaffold, and is more likely to yield truly biomimetic tissue biomimesis.

Synthetic hydrogels are also extensively used in tissue engineering. A well characterised material is polyethylene glycol (PEG). This material and other synthetic materials have the added benefit of being able to undergo extensive re-modelling processes prior to cell seeding to accurately control many of the material properties. These include methods such as photolithography and micromolding (Khademhosseini and Langer, 2007).

The requirements for engineering a cell-rich tissue or a matrix-rich tissue are considerably different. Cell-rich tissues are generally organs, and rather than conferring a mechanical role, tend to have limited strength. Compared to a matrix-rich tissue, that is, bone or tendon, these tissues play a crucial role in the mechanical integrity of the mammalian skeleton, and therefore the packing, orientation and density of the matrix proteins becomes paramount to this feature.

Both types of tissue have been successfully modelled using collagen scaffolds, including skin, skeletal muscle and tendon, but in terms of engineering the collagen protein, modelling of matrix-rich tissues has a greater impact.

© Woodhead Publishing Limited, 2013

20 Standardisation in cell and tissue engineering

2.1.1 Basic collagen hydrogel set-up

There are multiple methods of extraction of collagen from bulk tissues, and depending on the type of treatment, the collagen fibrils are rendered intact or a-telopeptic, that is, without the intact telopeptide ends. For 3D *in vitro* studies such disaggregated collagen needs to be reformed-undergo fibril-logenesis. This process *in vitro* entails bringing the collagen suspension to a neutral pH and increasing the temperature to 37°C.

For setting a collagen scaffold from native, intact collagen fibril solution extracted in acid, a basic ratio of 80% collagen solution, 10% 10× essential media solution (containing phenol red indicator), and 10% cell suspension should be used. However it remains paramount that the collagen solution is adequately neutralised before cells are added to prevent any adverse affects on cell viability.

An overview of this method is presented here:

1. Use a solution of extracted intact collagen fibrils in acetic acid. Notional value: 4 mL.
2. Add a 10 × essential media, containing 10 × (multi sign) the essential salt concentration and a phenol red indicator (used to observe the pH changes when the solution is neutralised). Notional value: 0.5 mL.
3. Neutralise this solution with 1 M NaOH till the pH is 7.3, or you observe a colour change from yellow to cirrus pink.
4. At this stage the collagen solution will begin to set (optimally at 37°C), so quickly add your cell suspension, mix gently and pour into the required mould for setting. Notional value of cell suspension: 0.5 mL.
5. The gel, with suspended cells will optimally set in a humidified incubator at 37°C and between 5% and 10% CO_2. This will take between 15 and 30 min.

This basic set-up for cellular collagen hydrogels has been used extensively to study cell behaviour and responses of various cell types in 3D (Cheema *et al.*, 2003; Grinnell and Petroll, 2010).

Standardisation of the amount of NaOH required to neutralise acetic acid is not possible due to the variability in the collagen source and the variation in amount of collagen extracted per batch.

The main parameters for setting collagen gels from acid solution are neutralisation to approximately neutral pH and bringing up to the correct temperature of 37°C. This makes controlling this process precisely difficult. For large-scale gel setting, the best alternative is to neutralise a large quantity of collagen (potentially 1000 mL or greater amounts) at 4°C, so that it doesn't set. This neutralised solution can then be stored, and as and when individual gels need to be set, the required amount can be removed and allowed to come up to 37°C.

© Woodhead Publishing Limited, 2013

2.2 Controlling the mechanical properties of collagen

Collagen hydrogels lack the necessary mechanical attributes to mimic tissues accurately. To optimally engineer a functional tissue using collagen as the major protein component, it will be critical to control the architectural features of collagen *in vitro*. Recently several strategies have been developed to tackle this issue, specifically to increase the collagen density, stiffness and strength. In the body many of these parameters will be controlled by cell involvement, which is not easily understood and therefore difficult to manipulate *in vitro* and as such novel physical approaches to control these parameters have been developed to control the major parameters dominating collagen architecture.

2.2.1 Controlling collagen density

Collagen type I is used for 3D culture of cells, providing a biomimetic environment in which to study cell behaviour (Grinnell and Petroll, 2010). Typically collagen scaffolds consist of collagen hydrogels, which as the name suggests, are mainly water. This generally means that the density of such scaffolds is inappropriate for modelling tissue matrix densities. These hydrogels are useful, however, in studying the interactions between cells and the surrounding matrix, as cells are able to remodel the matrix into which they have been seeded. They do so to orientate collagen fibrils, to control alignment of cells where strain is applied along an axis, and to encourage specific behaviours of cells in response to alignment, for example, fusion of single myoblasts to form multi-nucleated fibres (Cheema *et al.*, 2003). During this process the matrix is remodelled, and this tends to result in loss of water from the hydrogel, thus increasing the density of collagen.

Advances have been made to controllably increase the density of collagen scaffolds, starting with increasing the concentration of acid-collagen solutions through to expelling excess fluid using compression. Methods used to increase the concentration of collagen include evaporation, where acid-soluble collagen solution is poured into crystallising dishes, and following evaporation of water (in this case the solvent), concentrations of collagen increase (Bessea *et al.*, 2002). Methods employing reverse dialysis to concentrate collagen solutions have been employed (Knight *et al.*, 1998), as well as methods using continuous injection of low concentration collagen solutions into glass micro-chambers, to obtain dense collagen matrices (Mosser *et al.*, 2006). Both these methods have specific limitations, including the small volumes produced using dialysis and the formation of concentration gradients using the injection method, and recently a group have

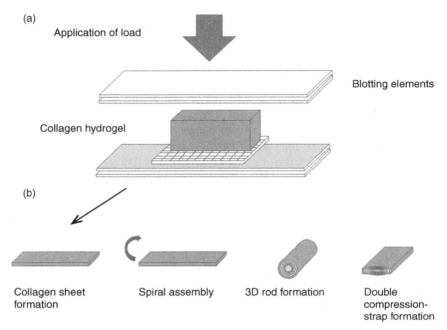

2.1 (a) Schematic of collagen hydrogel placed between layers of blotting elements, after which load is applied to plastically compress (PC) the gel. (b) Configurations of a PC gel, including simple sheet formation, spiral assembly and double compression to form dense 'straps' of collagen.

combined both dialysis and the injection method to form homogeneous dense fibrillar collagen matrices at high density (Wang *et al.*, 2011).

A recent application of controlled load to plastically compress (PC) standard collagen hydrogels to expel excess fluid and increase the collagen density has become particularly useful in the field of tissue engineering (Fig. 2.1) (Brown *et al.*, 2005). Application of this PC technology is very useful for *in vitro* tissue modelling as cell viability is retained in cells embedded within scaffolds undergoing PC, and the density increase (both for matrix and cells) is controllable and results in more biomimetic densities. At 11% collagen, standard PC techniques bring collagen scaffolds to *in vivo* levels of matrix density; however, the mechanical properties of such scaffolds still fall far short of those found in tissues. Further methods utilising PC technology have pushed these densities up to 30% and even higher (Abou-Neel *et al.*, 2006). But what is lacking is the specific architecture, and for this a multi-disciplinary approach to mimicking collagen architecture is required.

The basic method for application of compression (PC) relies upon placing a pre-set collagen hydrogel between layers of blotting paper and adding a set

Three-dimensional collagen biomatrix development and control   23

weight for a set period of time to essentially force water from the gel, whilst maintaining collagen integrity and cell viability. These weights have been established and standardised (Abou-Neel *et al.*, 2006) and are as follows:

- Single compression – application of 60 g weight (stress equivalent to approximately 1.4 kN/m^2) applied for 5 min under unconfined compression on the hyper-hydrated collagen gels.
- Double compression – further (second stage) application of 60 g (stress equivalent to approximately 22.6 kN/m^2) for 5 min, on an assembled 3D single-compressed construct.

This standardisation included an in-depth comparison of final collagen density, and scanning and transmission electron microscopy to ensure the same collagen structure was achieved when following this regime precisely.

Following plastic compression of collagen hydrogels, collagen sheets are formed which can either be cultured as they are, or rolled to give 3D rod formation (Fig. 2.1b). These 3D rods are in themselves good models to test gradient formation and will be discussed later. Further to this multiple compression regimes (specifically double compression) can be applied to further increase the collagen density of such scaffolds (Fig. 2.1b).

There are inherent changes in the scaffold properties as collagen density increases. As well as increasing strength and changing the stiffness of these scaffolds, these also include changes in the diffusion properties of these scaffolds. Following standard PC of collagen scaffolds, the diffusion of glucose, lactate and oxygen are all affected (Rong *et al.* 2006; Cheema *et al.*, 2011b). The diffusion coefficient for O_2 in PC collagen still falls within the range of some native tissues, including small intestinal submucosa, but what is important to note is that the presence of cells and formation of cell consumption gradients will compromise the diffusion of O_2 to the core of 3D assembled PC collagen constructs (Streeter and Cheema, 2011). Tissue engineering of cell-rich tissues using collagen scaffolds is therefore size-limited, unless adequate perfusion and/or vasculature can be engineered into the 3D construct.

2.2.2 Controlling collagen fibril diameter

Despite the importance of collagen fibril diameter to the material properties of tissues, our basic understanding of its control is poor. Control of fibril diameter is distinct from fibrillogenesis, which is the emergence of the tertiary collagen protein structure.

Fibril-modifying molecules, such as collagen types V and IX and proteoglycans such as decorin, are the main suggested mechanisms by which collagen fibril diameter is controlled *in vivo*, and these elements limit how large fibrils can grow in tissues such as cornea or tendon (Scott, 1984; Ezura *et al.*, 2000;

© Woodhead Publishing Limited, 2013

24 Standardisation in cell and tissue engineering

Ameye and Young, 2002). There is also an emerging understanding of how fibril diameters can be increased, by using mechanical forces to apply cyclical loading of collagen-containing gels, encouraging lateral fusion of fibrils, which is plausible when the quarter stagger patterns of the fibrils of collagen are in perfect register (Cheema *et al.*, 2007). This 'register' is most commonly identified as the banding pattern seen in transmission electron micrographs of native collagen fibrils. The need for banding pattern registration again lies in the short-range, non-covalent bonds presented between adjacent molecules, which drive fibril polymer formation. The bonding involved during this proposed fusion of fibrils is likely to be identical to the ionic and hydrogen bonding thought to stabilise the quarter stagger molecular packing in the original fibril. By application of 20% strain, followed by release of strain, up to three times per hour over a 48-h period, it is possible to significantly increase the diameter of collagen fibrils within plastically compressed collagen matrices (Cheema *et al.*, 2007).

Although this appears to be an engineering trick *in vitro*, it is highly likely that such mechanical forces occur in any new tissue *in vivo* under load. And importantly, the ability to control collagen fibril diameter without cells shows for the first time that mechanical forces *in vivo* may help determine fibril diameter and that cell-free engineering of native *collagen materials* is possible. Using technologies and strategies to manipulate and control fibril diameter will be critical to engineering collagen proteins for use as a suitable scaffold (Cheema *et al.*, 2011a).

2.2.3 Controlling collagen packing and orientation

The alignment of collagen fibrils in scaffolds is a critical parameter for control of architectural features. Without mimicking this alignment found in tissues, it is not possible to build a biomimetic tissue. Methods used *in vitro* to control alignment of collagen type I include magnetic alignment, interstitial directional fluid flow to control alignment and flow of collagen solution through microfluidic chambers during gelation (Elsdale and Bard, 1972; Girton *et al.*, 1999; Ng and Swartz, 2003; Lee *et al.*, 2006; Ng and Swartz, 2006; Guo and Kaufman, 2007). Elsdale and Bard were among the first groups able to align collagen. By simply setting a gel in a slanted chamber, and allowing interstitial fluid to flow downwards, collagen fibrils were observed to align along this fluid flow (Elsdale and Bard, 1972). This alignment strategy has recently been applied to collagen hydrogels using a similar fluid-flow mechanism, and yielded alignment along the axis of fluid flow, as well as an increase in total collagen density, measured up to 11%, showing that both fibril direction and density can be controlled using this fluid-flow mechanism (Kureshi *et al.*, 2010).

Tranquillo and colleagues have applied magnetic field to type I collagen scaffold, during gelation, and found that their collagen fibrils aligned along the plane in which the magnet was aligned (Girton *et al.*, 1999). Recently both

© Woodhead Publishing Limited, 2013

Three-dimensional collagen biomatrix development and control 25

the magnetic alignment and fluid-flow alignment methods have been applied together to controllably align collagen fibrils (Guo and Kaufman, 2007).

The methods described within this section will need to be critically used to generate the meso- and micro-scale architecture required to mimic tissues. There are limitations to each of these methods and further research into how to finely control collagen protein architecture is required. An example is how to control the bimodal distribution of fibril diameters within native tissues.

2.2.4 Controlling collagen cross-linking

Collagen is extensively cross-linked within the body. The mechanical strength of this collagen within tissues is brought about by the formation of intra- and intermolecular crosslinks. The main types are lysyl oxidase-mediated cross-links and non-enzymatic glycosylation of lysine and hydroxylysine residues. Cross-linking of collagen scaffolds has been achieved using, among others, chemical and physical methods. With the main chemical methods glutaraldehyde and formaldehyde are used, although both these chemicals render any embedded cells dead. Here the aldehyde reacts with the primary amine group of lysine and hydroxylysine of the collagen molecule. Due to the cytotoxicity of these chemicals, they are not viable options for cross-linking collagen in tissue-engineered constructs.

Recent advances have been made using the physical method of cross-linking, where ultra-violet (UV) irradiation of the photo-initiated compound riboflavin results in the formation of free radicals, which enhance cross-linking of collagen (Cheema *et al.*, 2011b). Riboflavin is well known for efficient photosensitised generation of singlet oxygen in oxygenated media and exhibits strong, broad absorption in the range 300–500 nm. About 0.25 mM riboflavin was added to 5 mL collagen scaffolds prior to compression. Once set, the collagen was irradiated at a distance of exactly 10 cm from a blue LED lamp for 10 min (Cheema *et al.*, 2011b). Application of this method has resulted in an increase in the break stress of plastically compressed collagen materials from 0.2 to 0.4 MPa where collagen scaffolds were subjected to photochemical cross-linking. A further benefit of using this method to cross-link collagen scaffolds is the limited effect it had on the viability of embedded cells. Although the increased cross-linking of collagen has resulted in stronger materials, the O_2 diffusion coefficient was found to decrease from 4.5×10^{-6} cm^2/s for a standard plastically compressed scaffold to 3.5×10^{-6} cm^2/s with photo-chemical cross-linking. By utilising techniques to alter or enhance specific material properties, we inevitably alter other material aspects, and a balanced understanding is needed to optimise the mechanical and material properties for any scaffold.

26 Standardisation in cell and tissue engineering

2.3 Architectural features: introducing elements of tissue complexity

Tissues are made of up of cells and protein matrix, but critical to the survival and functioning of any tissue is the presence of tissue- and nontissue-specific structures. Nontissue-specific structures include vascular structures, which supply all tissues with nutrients and neural structures, which provide a mechanism for sensory and motor control. Mimicking of such physical structures in collagen matrices is now possible with emerging new technologies.

For introduction of tubular physical structures, two main approaches have been studied. The first is incorporation of phosphate-based dissolving glass fibres into 3D scaffolds (Nazhat *et al.* 2007; Cheema *et al.*, 2010). As the name suggests, these fibres have a diameter of approximately 40 μm, and in studies to date, they dissolve in a period of about 24 h, although their chemical composition can be altered to change the rate of dissolution. These fibres can be incorporated into collagen matrices and following their dissolution leave continuous channels through which it is possible to track micro-bubble movement as well as delivery of molecules such as oxygen (Nazhat *et al.*, 2007). Incorporation of these fibres is straightforward, as they can simply be pressed into a setting collagen hydrogel, or placed onto a hydrogel, prior to compression, and following compression they will be embedded within the gel, and can be incorporated into the 3D configuration (Fig. 2.2). Following dissolution of fibres, channels remain open which can be seeded with cells specific for the structure required. In the case of vascular structures these would be endothelial cells (ECs).

Work has also been ongoing to build graduated structures into collagen, in the form of dissolvable conical fibres (Alekseeva *et al.*, 2011). These are smaller at one end, becoming gradually wider towards the other and providing a gradient along which cells can migrate and grow.

An alternative method to introduce physical features into collagen scaffolds is imprinting patterns into the surface, and then overlaying the imprinted side with further collagen scaffolds. This approach results in topographic features being incorporated into 3D structures, and cell responses can be monitored in such varying environments.

For complex biomimetic tissue-engineered constructs, composite scaffolds are being used to introduce multiple matrix proteins to accurately represent a native tissue's make-up. Where bone scaffolds are engineered, hydroxyapatite can be incorporated into collagen scaffolds, and where nerve conduits are developed, incorporation of fibronectin and other nerve matrix proteins is desirable (Brown *et al.*, 2005).

Engineering of vascular structures within tissue-engineered constructs is critical for larger-sized constructs, where nutrient and O_2 requirements by cells in the core are compromised. Development of vascular structures is

© Woodhead Publishing Limited, 2013

Three-dimensional collagen biomatrix development and control 27

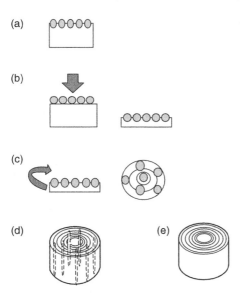

2.2 (a) Schematic representation of method to embed glass fibres into collagen hydrogels. (b) PC of collagen hydrogel with embedded fibres. (c) Spiral assembly of PC hydrogel, parallel to aligned fibres. (d) Assembly of 3D collagen construct, and the channels remaining following dissolution of the glass fibres. (e) Comparative collagen construct with no fibres.

primarily achieved by ECs and for maturation of these structures, the further addition of smooth muscle cells helps stabilise any structures formed, more similar to the composition of these structures within the body. A major body of work has focused on using matrigel as the 3D matrix component into which ECs are embedded, as spontaneous formation of tube-like structures occurs due to the presence of basement membrane proteins including laminin, collagen IV, heparin and entactin. Where cultures of aortic rings from rats have been embedded in matrigel cultures, sprouting of EC tube-like structures has been observed (Kubota *et al.*, 1988). However, the presence of these high concentrations of basement membrane proteins does interfere with the vessel growth, as luminal spaces were found to be significantly reduced in matrigel cultures compared to collagen hydrogels (Nicosia and Ottinetti, 1990). By adding individual components of basement membrane to collagen hydrogels, it is possible to optimise angiogenic sprouting in such cultures, as these proteins do stabilise the sprouting microvessels (Nicosia, 2009).

Addition of ECs into collagen hydrogels is currently a prerequisite for capillary structure formation; however, the addition of components like laminin and collagen IV into the collagen I matrix, as well as other basement

28   Standardisation in cell and tissue engineering

membrane components, is proving critical to the aggregation of ECs to form vessel-like structures.

2.3.1 Introducing cellular interfaces

Collagen models have been successfully engineered in which interfaces between cellular and acellular compartments, or different cellular compartments, are made. These models allow testing of specific cell–cell interactions in a spatially relevant manner. These interactions are incredibly complex to study *in vivo*, as deciphering and dissecting out specific functions of specific populations of cells is difficult. However, it is possible to study such signalling and cell–cell communication in 2D by utilising a biomimetic 3D model, which enables a more realistic scenario of cell behaviour and response to be achieved.

Examples of interface models using collagen include casting of collagen hydrogels with two different cell populations in different compartments segregated by a divider. The divider can then be removed after 2 min of setting to ensure integration of the interface without mixing the two cell populations (East *et al.*, 2011). Studies have been successfully conducted using this method to assess how a population of dissociated dorsal root ganglia can alter the phenotype of astrocytes by specific signalling (East *et al.*, 2011). In other examples of interface models using collagen scaffolds, pre-contracted collagen gels containing fibroblast cells are cast into freshly set acellular collagen gels. In these models, cell migration from cellular to acellular compartments of varying stiffness and density can be studied. This also provides a model to study interface adhesion, which mainly occurs via the actions of cellular remodelling at interfaces between tissues. For this the mechanical integration of the two compartments can be studied by simple pull-out tests which result in the determination of the extent of cellular remodelling of collagen at the interface (Marenzana *et al.*, 2007). The migration of cells from one compartment to another can be monitored using confocal microscopy and transferring images to appropriate imaging software packages.

2.3.2 Introduction of gradients in 3D matrices

Tissue growth *in vivo* is often dictated by the presence of gradients of growth factors and other molecules. Gradients of molecules, stiffness, cells and extracellular matrix all exist in native tissues, and to accurately mimic tissues, some of these gradients need to be recapitulated in our collagen biomatrix design.

Collagen is a native matrix protein found in the musculoskeletal tissues within the mammalian body. The diffusion coefficient of oxygen and glucose

Three-dimensional collagen biomatrix development and control   29

through dense (11% density) collagen scaffolds has been found to fall within the range of native tissues, including small intestinal submucosa (Cheema *et al.*, 2011b). When using these scaffolds to form 3D tissue constructs, cells embedded in the 3D scaffold are instrumental in the formation of consumption gradients for most molecules, including oxygen. As there is no inherent vascular system within tissue constructs, these consumption gradients result in exposure of oxygen being mainly dependent upon the spatial location of specific populations of cells (Cheema *et al.*, 2008; Fig. 2.3). An example would be cells on the surface of a 3D collagen scaffold being exposed to much higher levels of oxygen compared to cells within the core. In response to this gradient of oxygen, cell behaviour is simultaneously also affected. An example would be the increased stimulation of angiogenic factors by cells within the core of 3D scaffolds compared to the surface (Cheema *et al.*, 2008). This system is easily optimised by altering cell density so that O_2 levels in the core never reach pathological hypoxic levels, but remain within physiological hypoxia to ensure cell survival, even though angiogenic signalling is maximally up-regulated. Levels of angiogenic signalling gradients of molecules can be controlled by altering the cell-seeding density parameter.

As well as this natural formation of consumption gradients, 3D collagen constructs can be designed so that cell-generated molecules are formed at specific locations (Fig. 2.3a). Using the same premise that high cell-density scaffolds result in the formation of oxygen consumption gradients, or local cell-mediated hypoxia, we can position depots of these cell scaffolds within collagen matrices to engineer cell-generated molecules in specific locations (Fig. 2.3b) (Hadjipanayi *et al.*, 2010). In a recent study cell-mediated hypoxia was engineered using human dermal fibroblasts (HDFs) to generate a local population of hypoxia-induced signalling (HIS) cells in 3D collagen scaffold depots. HIS cell depots released angiogenic factors, in a gradient fashion, which induced directional EC migration and tubule formation in a spatially defined assay system. Hence spatially positioned local hypoxic stimuli using defined cell-collagen depots can be introduced into larger and more complex 3D collagen scaffolds to form gradients of molecules.

Cell behaviour is affected not only by molecules and growth factors, but also by physical geometric factors, including stiffness. Stiffness gradients can also be introduced to 3D collagen scaffolds using easily reproducible methods (Hadjipanayi *et al.*, 2009b). The main technique for creation of a stiffness gradient within a collagen scaffold involves casting a collagen hydrogel in the shape of a wedge, so there is more collagen content at one end compared to the other. Following even, continuous plastic compression of the entire construct, one end contains more collagen, and is hence much stiffer than the opposite end (Hadjipanayi *et al.*, 2009b). The increase in stiffness of collagen matrices has been found to enhance specific cell behaviours, including cell migration and cell proliferation (Hadjipanayi *et al.*, 2009a, 2009b).

(a)

Cell seeding of 3D
collagen scaffold

Formation of O_2 gradient –
resulting in angiogenic
signalling by core cells

Unfurl constructs to
dissect cell populations
exposed to varying O_2

(b)

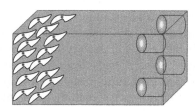

Endothelial   Angiogenic
cells   depots

2.3 (a) Schematic of O_2 gradient formation within a cell-seeded 3D collagen construct, and the method to dissect out the core, mid and outer sections of the gel. Cells trapped within these regions can then be assessed for viability using live/dead staining, or gene upregulation following RNA extraction from cells. (b) Positioning of angiogenic depots at one end of a larger collagen hydrogel scaffold, which release factors into the matrix to attract endothelial cells. Different cells can then be embedded at the opposite end, and migration of these cells towards angiogenic signalling can be monitored.

2.4 Future trends

The formation of tissue *in vivo* is an iterative process where cells lay down matrix proteins, re-model this matrix and then continue to lay down further matrix. In a similar way, through the actions of exercise, various tissues are re-modelled by cells to adapt to the changing mechanical stresses to which they are subjected. Re-creating this scenario *in vitro* by relying on cell remodelling is very time-consuming and not easy to control. By using engineering tricks we can create more biomimetic material properties of collagen scaffolds, and following cell addition, still rely on aspects of cellular remodelling to create viable and biomimetic tissues within reasonable timeframes.

As collagen I is the dominant protein of the majority of tissues of the body, the architecture and packing of this protein need to be controlled if

Three-dimensional collagen biomatrix development and control   31

we are to successfully engineer tissues *in vitro*. Many collagen scaffold materials are currently available for use *in vivo* as replacement skin, tissue fillers and artificial vascular structures, but the best of these have relied upon retaining the original architecture of the collagen from tissues.

As our understanding of how this protein is modified and packed *in vivo* by cells, mechanical load and other matrix proteins progresses, we can systematically apply these procedures to predictably control collagen architecture, matrix composition within scaffolds, cellular behaviour and cellular remodelling of extra-cellular protein. This not only has implications for using *in vitro* models as tissue test-bed equivalents, but will more likely result in biomimetic tissue-engineered constructs which may have clinical benefit.

2.5   References

Abou-Neel, E.A., Cheema, U., Knowles, J.C., Brown, R.A., and Nazhat, S.N. 2006. Use of multiple unconfined compression for fine control of collagen gel scaffold density and mechanical properties. *Soft Matter*, **2**, 986–992.

Alekseeva, T., Neel, E.A., Knowles, J.C., and Brown, R.A. 2011. Development of conical soluble phosphate glass fibers for directional tissue growth. *J Biomater Appl*, **20**, 274–280.

Ameye, L. and Young, M.F. 2002. Mice deficient in small leucine-rich proteoglycans: novel *in vivo* models for osteoporosis, osteoarthritis, Ehlers-Danlos syndrome, muscular dystrophy, and corneal diseases. *Glycobiology*, **12**(9), 107R–116R.

Bessea, L., Coulomb, B., Lebreton-Decoster, C., and Giraud-Guille, M.M. 2002. Production of ordered collagen matrices for three-dimensional cell culture. *Biomaterials*, **23**(1), 27–36.

Brown, R.A., Wiseman, M., Chuo, C.B., Cheema, U., and Nazhat, S.N. 2005. Ultrarapid engineering of biomimetic materials and tissues: fabrication of nano- and microstructures by plastic compression. *Adv Funct Mat*, **15**(11), 1762–1770.

Cheema, U., Alekseeva, T., Abou-Neel, E.A., and Brown, R.A. 2010. Switching off angiogenic signalling: creating channelled constructs for adequate oxygen delivery in tissue engineered constructs. *Eur Cell Mater*, **20**, 274–280.

Cheema, U., Ananta, M., and Mudera, V. 2011a. Collagen: applications of a natural polymer in regenerative medicine. In *Regenerative Medicine and Tissue Engineering*, First edition. D. Eberli, ed., Rijeka: Intech, pp. 287–300.

Cheema, U., Brown, R.A., Alp, B., and Macrobert, A.J. 2008. Spatially defined oxygen gradients and vascular endothelial growth factor expression in an engineered 3D cell model. *Cell Mol Life Sci*, **65**(1), 177–186.

Cheema, U., Chuo, C.B., Sarathchandra, P., Nazhat, S.N., and Brown, R.A. 2007. Engineering functional collagen scaffolds: cyclical loading increases material strength and fibril aggregation. *Adv Funct Mat*, **17**, 2426–2431.

Cheema, U., Rong, Z., Kirresh, O., Macrobert, A.J., Vadgama, P., and Brown, R.A. 2011b. Oxygen diffusion through collagen scaffolds at defined densities: implications for cell survival in tissue models. *J Tissue Eng Regen Med*, **6**(1), 77–84.

32   Standardisation in cell and tissue engineering

Cheema, U., Yang, S.Y., Mudera, V., Goldspink, G., and Brown, R.A. 2003. 3-D *In Vitro* model of early skeletal muscle development. *Cell Motil Cytoskel*, **54**, 226–236.

East, E., Golding, J.P., and Phillips, J.B. 2011. Development of an integrated collagen gel system for studying cellular interfaces following spinal cord injury. *Eur Cell Mater*, **22**(2), 3.

Elsdale, T. and Bard, J. 1972. Collagen substrata for studies on cell behavior. *J Cell Biol*, **54**(3), 626–637.

Ezura, Y., Chakravarti, S., Oldberg, A., Chervoneva, I., and Birk, D.E. 2000. Differential expression of lumican and fibromodulin regulate collagen fibrillogenesis in developing mouse tendons. *J Cell Biol*, **151**(4), 779–788.

Girton, T.S., Dubey, N., and Tranquillo, R.T. 1999. Magnetic-induced alignment of collagen fibrils in tissue equivalents. *Methods Mol Med*, **18**, 67–73.

Grinnell, F. and Petroll, W.M. 2010. Cell motility and mechanics in three-dimensional collagen matrices. *Ann Rev Cell Dev Biol*, **26**, 335–361.

Guo, C. and Kaufman, L.J. 2007. Flow and magnetic field induced collagen alignment. *Biomaterials*, **28**(6), 1105–1114.

Hadjipanayi, E., Brown, R.A., Mudera, V., Deng, D., Liu, W., and Cheema, U. 2010. Controlling physiological angiogenesis by hypoxia-induced signaling. *J Control Release*, **146**(3), 309–317.

Hadjipanayi, E., Mudera, V., and Brown, R.A. 2009a. Close dependence of fibroblast proliferation on collagen scaffold matrix stiffness. *J Tissue Eng Regen Med*, **3**(2), 77–84.

Hadjipanayi, E., Mudera, V., and Brown, R.A. 2009b. Guiding cell migration in 3D: a collagen matrix with graded directional stiffness. *Cell Motil Cytoskel*, **66**(3), 121–128.

Khademhosseini, A. and Langer, R. 2007. Microengineered hydrogels for tissue engineering. *Biomaterials*, **28**(34), 5087–5092.

Knight, D.P., Nash, L., Hu, X.W., Haffegee, J., and Ho, M.W. 1998. *In vitro* formation by reverse dialysis of collagen gels containing highly oriented arrays of fibrils. *J Biomed Mater Res*, **41**(2), 185–191.

Kubota, Y., Kleinman, H.K., Martin, G.R., and Lawley, T.J. 1988. Role of laminin and basement membrane in the morphological differentiation of human endothelial cells into capillary-like structures. *J Cell Biol*, **107**(4), 1589–1598.

Kureshi, A., Cheema, U., Alekseeva, T.A., Cambrey, A., and Brown, R.A. 2010. Alignment hierarchies: engineering architecture from the nanometre to the micrometre scale. *J R Soc Interface*, **6**, S107–S116.

Lee, P., Lin, R., Moon, J., and Lee, L.P. 2006. Microfluidic alignment of collagen fibers for *in vitro* cell culture. *Biomed Microdevices*, **8**, 35–41.

Marenzana, M., Kelly, D.J., Prendergast, P.J., and Brown, R.A. 2007. A collagen-based interface construct for the assessment of cell-dependent mechanical integration of tissue surfaces. *Cell Tissue Res*, **327**(2), 293–300.

Mosser, G., Anglo, A., Helary, C., Bouligand, Y., and Giraud-Guille, M.M. 2006. Dense tissue-like collagen matrices formed in cell-free conditions. *Matrix Biol*, **25**(1), 3–13.

Nazhat, S.N., Neel, E.A., Kidane, A., Ahmed, I., Hope, C., Kershaw, M., Lee, P.D., Stride, E., Saffari, N., Knowles, J.C., and Brown, R.A. 2007. Controlled micro-

Three-dimensional collagen biomatrix development and control   33

channelling in dense collagen scaffolds by soluble phosphate glass fibers. *Biomacromolecules*, **8**(2), 543–551.

Ng, C.P. and Swartz, M.A. 2003. Fibroblast alignment under interstitial fluid flow using a novel 3-D tissue culture model. *Am J Physiol-Heart C*, **284**, H1771–H1777.

Ng, C.P. and Swartz, M.A. 2006. Mechanisms of interstitial flow-induced remodeling of fibroblast–collagen cultures. *Ann Biomed Eng*, **34**, 446–454.

Nicosia, R.F. 2009. The aortic ring model of angiogenesis: a quarter century of search and discovery. *J Cell Mol Med*, **13**(10), 4113–4136.

Nicosia, R.F. and Ottinetti, A. 1990. Modulation of microvascular growth and morphogenesis by reconstituted basement membrane gel in three-dimensional cultures of rat aorta: a comparative study of angiogenesis in matrigel, collagen, fibrin, and plasma clot. *In Vitro Cell Dev Biol*, **26**(2), 119–128.

Rong, Z., Cheema, U., and Vadgama, P. 2006. Needle enzyme electrode based glucose diffusive transport measurement in a collagen gel and validation of a simulation model. *Analyst*, **131**(7), 816–821.

Rowley, J.A., Madlambayan, G., and Mooney, D.J. 1999. Alginate hydrogels as synthetic extracellular matrix materials. *Biomaterials*, **20**(1), 45–53.

Scott, J.E. 1984. The periphery of the developing collagen fibril. *Biochem J*, **195**, 229–233.

Streeter, I. and Cheema, U. 2011. Oxygen consumption rate of cells in 3D culture: the use of experiment and simulation to measure kinetic parameters and optimise culture conditions. *Analyst*, **136**(19), 4013–4019.

Wang, Y., Silvent, J., Robin, M., Babonneau, F., Meddahi-Pelle, A., Nassif, N., and Guille, M.M.G. 2011. Controlled collagen assembly to build dense tissue-like materials for tissue engineering. *Soft Matter*, **7**, 9659–9664.

3
Two- and three-dimensional tissue culture bioprocessing methods for soft tissue engineering

M. J. ELLIS, University of Bath, UK

DOI: 10.1533/9780857098726.1.34

Abstract: This chapter outlines bioreactor design for soft tissue, using a bioprocessing approach. Bioreactor configurations are reviewed, and as bioreactor performance is intimately related to the scaffold, materials and scaffold architecture are described. Mass transfer considerations, bioreactor data collection and how to take a bioprocessing approach to bioreactor design are described, with an example of liver tissue engineering.

Key words: bioreactor, tissue engineering, scaffold, bioprocessing, biochemical engineering, regenerative medicine, vascularisation.

3.1 Introduction

This chapter will outline how to design a bioreactor system for tissue engineering, by considering all components. It will introduce the different bioreactor configurations and provide some examples where they are already in use, which will provide a good summary for engineers and a good introduction for those new to engineering. As bioreactor performance is intimately related to the scaffold, materials and scaffold architecture will be described, as well as the fabrication techniques that are used to make different scaffolds. Cell–scaffold interactions, surface modification and controlled release will be briefly covered. An introduction to mass transfer and fluid dynamics is given, with examples on hollow fibre bioreactors, followed by a description of measurements necessary to accurately assess and model a bioreactor. This section follows the standard chemical engineering approach plus some new advances in the field of tissue engineering. Finally, an introduction to bioprocessing brings together all the previous sections and includes an example of how a bioreactor for an engineered liver regeneration construct could be designed. While this chapter is focused on soft tissue, that is, non-loadbearing tissue, the principles can be applied to any tissue type. As such, this chapter will provide a valuable introductory resource for bioreactor design and operation, from selecting the building blocks to process design.

3.2 Bioreactor configurations

Reactor is the name given to a unit, a piece of equipment that has an input of raw materials, say A, B and C, and an output of products D, E and F. A reaction has taken place and in the case of bioreactors, a micro-organism was involved in the conversion. So, in the context of tissue engineering, mammalian cells are the micro-organism, media provides the raw materials (and perhaps the scaffold if it is used as a delivery vehicle for the controlled release of growth factors and drugs), and more cells or differentiated cells, and metabolites are the products. Any device that is used to culture the cells can be called a bioreactor, whether it is a T-flask, 96-well plate, Techne flask©, or one of the other configurations of reactor designs. There are six main reactor configurations and all reactors and bioreactor designs fall into one of these categories, which are outlined below. Media is taken to be the raw materials, metabolites both useful and waste are taken to be the products.

3.2.1 Batch reactors

Batch reactors are filled with the media and nothing is added or removed until the end of the reaction. A semi-batch reactor will have some addition and/or removal during the course of the reaction; a T-flask is a semi-batch reactor if media is changed between passages. Batch reactors can be operated as static or mixed; a T-flask is static (Fig. 3.1) and a stirrer flask, such as a Techne® Flask (Fig. 3.2) is mixed. Mixing helps remove concentration gradients and provides shear. Static batch are poorly mixed, do not induce shear, and rely on diffusion for mass transfer. Mixing provides a homogeneous environment and shear, while mass transfer will be by both convection and diffusion; however, as the media is depleted and products build up the optimum culture conditions are soon lost.

3.2.2 Continuous stirred tank reactors (CSTR)

A continuous stirred tank reactor (CSTR) is mixed and has a continuous feed of media and an equal rate of removal of products. At equilibrium the conditions in the CSTR are homogeneous and the outlet conditions are taken to be the same as the bulk, which is a mixture of media and products. Techne® Flasks have the facility to operate as a CSTR. Mass transfer will have convective and diffusional components.

3.1 A T75 cell culture flask. This is an example of a semi-batch bioreactor as media is usually changed every 3 days; T75 refers to a T-flask with a culture surface of 75 cm^2.

3.2.3 Flat bed reactors

In a flat bed, the media flows over the cells on the reactor bed and is utilised while products are released into the media. Concentration gradients exist along the length of the flat bed as media concentration decreases and product concentration increases. Shear is induced by the continuous flow of media. Concentration and shear are coupled as changes in flow rate to control the concentration gradients will alter the shear stress. Flat beds are used in, for example, *in vitro* liver models (Allen and Bhatia, 2003), and microreactors (Fig. 3.3). Mass transfer will have convective and diffusional components.

3.2.4 Packed bed bioreactors

Packed bed bioreactors contain packing, which is either particulate (Fig. 3.4a) or a porous block. Cells adhere to the surfaces of the packing, that is, the scaffold, or can be encapsulated within it, and media flows through the usually tortuous channels. Careful design of the scaffold will provide a high surface area to volume ratio and so allow an efficient system design. Poor design will result in 'dead zones' where no cells are able to grow either due to poor mass transfer or localised high shear stresses. Packed beds can be orientated vertically with up-flow or down-flow (a trickle bed bioreactor) of media, or horizontally. Most tissue engineering bioreactors are based on this configuration, and use a range of packing as the scaffold, for example the extensive work by Shakesheff and co-workers using a poly(lactide-co-glycolide) foam produced using supercritical CO$_2$ (Yang *et al.*, 2001). Mass transfer will have convective and diffusional components.

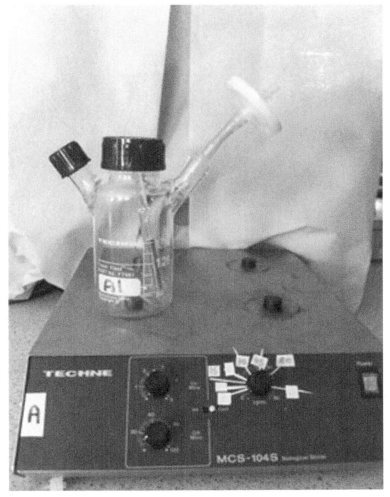

3.2 A Techne® Flask. This is an example of a mixed batch reactor. The stirrer is magnetic and the base on which it stands contains a magnet to activate the stirrer. An HEPA (high-efficiency particulate air) filter has been fitted in the right-hand port to allow gaseous exchange with the environment.

3.2.5 Fluidised bed bioreactors

Fluidised bed bioreactors contain particulate scaffolds with cells, or cell spheroids (Fig. 3.4b and 3.4d). How the cells are contained in or on the scaffold is of importance to the specific application; cells can be adhered on or in the particles, or encapsulated. The bioreactor is vertical with up-flow (although down-flow can be used prior or after fluidisation in a trickle bed

38   Standardisation in cell and tissue engineering

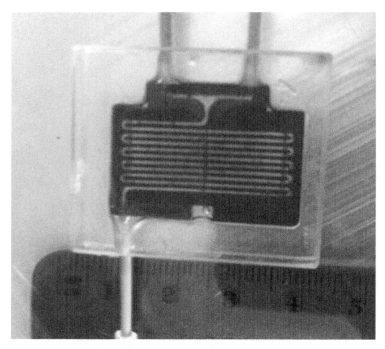

3.3 A microreactor. This is an example of a flat bed bioreactor. This in-house design has channels with a flat base on which cells are cultured. Photo courtesy of Fernando Acosta, Department of Chemical Engineering, University of Bath.

configuration if desired), and the flow rate is such that the buoyancy of the particles is overcome and the fluid lifts the particles so they circulate in the column. Bed expansion occurs prior to full fluidisation, and it has been shown that cell culture can be successfully carried out in an expanded bed set-up (Storm *et al.*, 2010). Expanded beds and fluidised beds provide a high surface area to volume ratio, and because each particle is stand-alone and media passes around it, the mass transfer problems found with a packed bed are eliminated. Shear is induced by the flow and the drawback with this design is that the shear may be too great for the cells to survive or remain attached to the scaffold. Mass transfer will have convective and diffusional components.

3.2.6 Membrane bioreactors

Membrane bioreactors can have flat sheet, hollow fibre, tubular or spiral-wound membranes; examples of hollow fibre membrane bioreactor modules are shown in Fig. 3.5. Membranes are semi-permeable barriers and

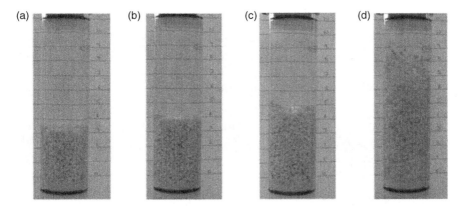

3.4 Packed bed and fluidised bed bioreactors. This sequence of photographs shows hydroxyapatite particles in a column containing water, being used as a preliminary study to assess the pump requirements. As the flow rate is increased the set-up changes from (a) a packed bed where the particles are settled on the bottom of the column, to (b) an expanded bed where the particles are just separated by the up-flow of water, to (c and d) a fluidised bed. Photo courtesy of Dr Ian Benzeval, Department of Chemical Engineering, University of Bath.

are found throughout the body to control the movement of nutrients and waste products in the various organs, for example blood vessel walls and the kidney. Therefore membrane bioreactors offer the potential to closely replicate the physiological environment. Media is fed on one side of the membrane and passes across the membrane wall to the cells on the other side. Shear can be induced on the cells independently to the main media feed by a secondary flow on the shell-side. Waste products can also be removed via side ports. Careful design allows a very high surface area to volume ratio. Membrane bioreactors have been used to replicate vascularisation in tissues such as bone (Ellis and Chaudhuri, 2007; Morgan *et al.*, 2009) and liver (Curcio *et al.*, 2005; Lu *et al.*, 2005) and the filter system in lung (Grek *et al.*, 2009) and kidney (Oo *et al.*, 2011), and organ assist devices such as the Extracorporeal Liver Assist Device (ELAD) system by Vital Therapies Inc. Tubular membranes have been used for nerve guidance (Oudega *et al.*, 2001). Mass transfer will have convective and diffusional components.

3.3 Selecting scaffold materials and architectures for your bioreactor

A chapter on bioreactors would not be complete without a discussion on scaffolds. There are many good texts, articles and reviews on the subject, and so here is an overview with an emphasis on systematic design.

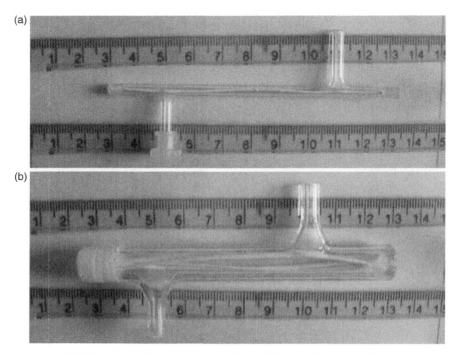

3.5 Membrane bioreactor modules. (a) Single-fibre module, which represents a single sub-unit of a hollow fibre bioreactor. (b) Small-scale hollow fibre bioreactor containing fibres with a total surface area of 25 cm^2.

3.3.1 Scaffold materials

Materials for scaffolds are polymers, which can be synthetic or natural, hydrogels, ceramics and metals (Ellis *et al.*, 2008). Tissue engineers are fortunate that a significant body of work has been carried out, and continues to grow, in the design and development of materials that can be used. As our understanding of cell–material interactions, and the effects of these on cell behaviour, develops we get ever closer to replicating the *in vivo* environment. Natural polymers such as collagen provide a matrix most similar to that found *in vivo* as they are sourced from decellularised tissue. Badylak (2004) reviews xenogenic scaffolds, of which porcine-derived is the most common. Synthetic polymers are highly versatile, and tend to have less batch variation compared to animal-derived material. Gels are once again receiving considerable focus, particularly now that injectable scaffolds are popular for soft tissues including nerve, adipose and heart regeneration (Liu *et al.*, 2012; Macaya and Spector, 2012; Wu *et al.*, 2012). Ceramics and metals are suited to hard tissue engineering and will not be covered here. A great advantage of biodegradable polymeric and gel scaffolds is their ability to allow controlled release of chemicals. This is well established in controlled drug delivery and is now finding its place in tissue engineering and regenerative medicine, so scaffolds are not just delivering cells but active biological molecules as well; the reader is directed to (Saltzman, 2001; Nair and Laurencin, 2006).

3.3.2 Cell attachment and surface modification

When selecting and developing a scaffold it is necessary to appreciate the parameters involved in the molecular design and the resulting chemical and physical properties; understand cell binding to extracellular matrix proteins, commonly RDG-integrin binding; and recognise a range of surface modification methods and if and when to apply these based on the selected material and cell type. Lactide-based polymers are usually hydrophobic (non-polar), with a contact angle greater than 65° (Ellis *et al.*, 2008) while proteins are usually hydrophilic (polar). Many cells attach better to positively charged moeities since the cell surface has an overall negative charge, due to sialic acids(Varki, 1997). TCPS (tissue culture polystyrene) is treated so there are available oxygen-containing groups on the surface and serum proteins in cell culture media bind to TCPS surface. Cells bind to the serum proteins via integrins. Therefore scaffolds should provide at least a good attachment for serum proteins as TCPS, and ideally provide attachment sites without the need for serum. Extracellular matrix proteins of common interest are collagen, fibronectin, vitronectin, laminin, and proteoglycans (Ellis *et al.*, 2008). Vitronectin is the mediator for cell attachment to TCPS, and fibronectin, laminin and collagen (particularly Collagen I) are commonly used to coat scaffolds. These proteins are found in serum, although not in specified amounts; as such it is sensible to batch test serum and reserve it for a set of studies. It should be noted that attaching the RGD sequence may not be enough as binding involves the synergy site as well as the amino acid sequence itself (Redick *et al.*, 2000; Perlin *et al.*, 2008). Other chemical, physical and mechanical factors affect cell attachment, proliferation and function, and current understanding on this is growing rapidly. It has been known for some time now that topography has a significant effect on cell attachment and migration (Dalby *et al.*, 2003; Biggs *et al.*, 2010; Wu *et al.*, 2010; McNamara *et al.*, 2012; Fig. 3.6), and that differentiation can be directed by varying material stiffness (Discher *et al.*, 2005). This has been demonstrated for directed mesenchymal stem cell differentiation along different pathways (Pek *et al.*, 2010; Murphy *et al.*, 2011), and for cornea (Jones *et al.*, 2012). An extension of this is that the bond strength can also affect attachment, migration and also differentiation. Catalyst residue affects cell

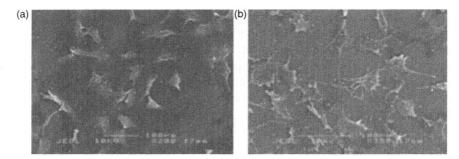

3.6 MG63 cells on poly (lactide-co-glycolide) flat sheet scaffolds with different topography. The osteosarcoma cell line MG63 presents different morphology on different surfaces: (a) smooth, (b) honeycomb-structured. The cells show more protrusions on the honeycomb-structured surface compared to the smooth surface.

behaviour, for example, tin is toxic and we have shown zirconium gives preferential cell culture results for MG63 cells.

3.3.3 Scaffold architecture

The scaffold architecture is usually defined by the tissue properties and the end use, for example, how it is going to be implanted. Scaffold material and surface modification will influence the scaffold architecture that can be used, and should be kept in mind when choosing the architecture; polymers are the most versatile when it comes to the range of architectures (Ellis *et al.*, 2008). The selection of bioreactor configuration is dependent on the scaffold architecture, a summary of which is shown in Table 3.1.

A film or sheet has been shown to be successful for skin (Zhu *et al.*, 2005, 2008) and cornea (Lawrence *et al.*, 2009; Fagerholm *et al.*, 2010; Bray *et al.*, 2012); porous blocks are most extensively used for bone but can be applied to a range of soft tissues (Yang *et al.*, 2001; Jones *et al.*, 2010; Lim and Park, 2011; Salerno *et al.*, 2011); microspheres as stem cell carriers (Newman and McBurney, 2004) can also be used for a range of tissues including adipose (Kimura *et al.*, 2003; Yu *et al.*, 2005); fibres (solid and distinguished from hollow fibres) are very versatile and can be formed using electrospinning to make random-fibre porous mats for skin (Zhu *et al.*, 2008) or aligned fibres for nerve guidance (Yang *et al.*, 2005) or made into tubes for vascular engineering (Soffer *et al.*, 2008); tubes have also been used for nerve (Oudega *et al.*, 2001); and hollow fibres are a popular choice to represent vasculature, termed 'pseudovascularisation', and have proven successful for liver in particular (Curcio *et al.*, 2005; Lu *et al.*,

2- and 3-D tissue culture bioprocessing methods 43

Table 3.1 Bioreactor types and scaffold architectures (crosses indicate which scaffold architectures can be used, in principle, in the different bioreactor types)

Scaffold architecture	Batch	Flat bed	CSTR*	Packed bed	Fluidised bed	Membrane
Film	x	x				
Porous sheet	x	x				
Porous block/ sponge	x			x		
Channelled block/ sponge	x			x		
Fibres	x	x		x		x
Hollow fibres/tubes	x					x
Microspheres/ particles	x		x	x	x	

*CSTR is 'continuous stirred tank reactor'.

2005). The range of scaffold architectures and how they are dependent on material and fabrication process is summarised in Ellis *et al.* (2008).

3.4 Mass transfer in tissue engineering bioreactors

Bioreactors deliver media to cells and remove products. This mass transfer will occur as a function of the fluid dynamics in the system and occurs at the macro-, micro- and nano-scale. Macro-scale flow is blood flow through vessels, or bulk media flow in a bioreactor, at the cm-scale, and convection is usually dominant over diffusion. Micro-scale flow is interstitial flow in tissue to the cell, or movement of media from the bulk flow to the cell in a bioreactor, at the micrometer level and having convective and diffusive components, each of which can dominate. At the nano-scale, molecules move across the cell membrane and through the cell, at the nanometer level. Good texts for the macro- and micro-scale and transport into and out of cells include Doran (1997) and Saltzman (2001). Intercellular transport, that is, signalling, is vital for cell-culture analysis and for understanding how to manipulate the cell environment, and the reader is referred to one of the many books on signalling such as the superb online book 'Cell Signalling Biology' by Professor Sir Michael J Berridge (Berridge, 2009), plus journal papers, as the field is moving very quickly.

For bioreactor design, operating and monitoring, oxygen is a good molecule to study as it is vital to cell processes, is often the limiting nutrient, and is relatively cheap and easy to monitor. The partial pressure of oxygen (pO_2) varies throughout the body, is different in different tissues and can vary within a tissue. It is known to play an important role in embryonic

© Woodhead Publishing Limited, 2013

44 Standardisation in cell and tissue engineering

development, and will affect cell behaviour and phenotype, including zonation in liver (gradient of metabolic functions along a liver lobule), proliferation, and differentiation (Wion *et al.*, 2009; Bambrick *et al.*, 2011). To replicate physiological conditions we need to know the O_2 profile spatially and temporally throughout the culture system. Oxygen gradients will exist as a function of oxygen uptake and fluid dynamics. In practice this means that the concentration of the oxygen in the media is not the concentration the cell experiences. Pericellular pO_2 is a function of cell type, cell density, and cell cycle phase (Bambrick *et al.*, 2011), and the fluid dynamic profile. Doran (1997) introduces the calculation of O_2 transfer for a well-mixed sparged fermenter (i.e., gas, in this case air, is pumped up through the fermenter), which provides a good starting point for flask culture and can be modified for perfused culture.

Knowing the oxygen requirements and ensuring all cells receive at least this minimum supply is key to bioreactor design. Take a hollow fibre bioreactor; it is known that the point of minimal oxygen concentration (c_{min}) will be at the further distance from the inlet. Assuming Michalis–Menton kinetics and obtaining oxygen requirement data from literature for a range of tissues, Shipley *et al.* (2011) were able to prescribe design equations for the lumen length, distance between fibres, and media flow rate. This paper also neatly highlights the appropriate use of analytical and numerical modelling since each were seen to be suitable under different conditions. Specifically when $c_{min} \gg K_m$ (where K_m is the half-maximal oxygen concentration) an analytical approach is appropriate; when $c_{min} >/> K_m$ a numerical approach is appropriate. Mathematical modelling is essential for the detailed design of bioreactors but care must be taken to approach this in the most suitable way and by making sensible assumptions based on a sound knowledge of the biological system, the desired outcome and the bioreactor set-up.

The modelling of oxygen gradients has been taken one step further in an attempt to model and therefore design a zonated liver bioreactor, in a flat bed by Allen and Bhatia (2003) and a hollow fibre bioreactor by Davidson *et al.* (2012). Metabolic zonation is regulated by oxygen gradient, hormones and extracellular matrix, and here the focus was on oxygen. Based on known parameters for bioartificial livers, the properties of the hollow fibre bioreactor, Davidson *et al.* (2012) showed that geometry and flow rates can be found to provide a prescribed oxygen gradient, and furthermore each zone length can within reason be specified in relation to the other two.

Once the oxygen requirements are known and the geometry defined, the operating conditions of the bioreactor must be set. Using Equation [3.1] (Shipley *et al.*, 2010), if the bioreactor configuration is known (A and B) and the desired permeate (the phase that crosses the membrane wall) to retentate (the phase that remains on the feed-side, in this case in the lumen) ratio

© Woodhead Publishing Limited, 2013

2- and 3-D tissue culture bioprocessing methods   45

Table 3.2 Physical, chemical and biological parameters that should be measured during bioreactor operation

Physical	Chemical	Biological
Flowrates	Oxygen concentration	Viability
Pressures	Nutrient and other media component concentrations	Metabolic activity
Temperatures	Other additive concentrations (drugs/ growth factors/chemicals)	Cell number
	Metabolite concentrations	Cell structure Gene expression

known (c), then the pressure drop between the lumen inlet (P_1) and outlet (P_0) can be found for a given volumetric flow rate $(Q_{1,in})$ or *vice versa*.

$$P_1 - P_0 = Q_{1,in}(Ac + B) \qquad [3.1]$$

Accurate depiction of the fluid dynamics and mass transfer is essential for successful bioreactor operation. A good mathematical model is based on good experimental data, and will allow for modification as more and better data is obtained. Hence, mathematical modelling and experimental design is an iterative process.

3.5   Important parameters and taking measurements of bioreactor cultures

A bioreactor design is only as good as the data used to design it. The more you have and the more accurate it is, the better the final design will be, in terms of effectiveness and efficiency. Measurements need to be taken during start-up until the system equilibrates, then for the duration of the operation. A list of physical, chemical and biological measurements required is shown in Table 3.2. Common measurements can be taking using the following tools.

For quantitative and semi-quantitative measurements the following can be used:

- In-line probes for flow rate, pressure and temperature.
- Cell function kits such as picogreen® for DNA quantification (cell number) and MTT for viability.
- Media analysis kits such as those by Megazyme for D-glucose, L-glutamine/ammonia and L-lactic acid.
- Western blotting for protein expression.
- Polymerase chain reaction (PCR) for gene expression.

46 Standardisation in cell and tissue engineering

- High performance liquid chromatography (HPLC) to quantify media components.
- Flow cytometry to analyse cell phenotype by cell-surface antigens, and relative DNA quantification.
- Atomic force microscopy to measure cell-surface bond strength.

For visualisation the following can be used:

- Histology stains to analyse cell anatomy.
- Scanning electron microscopy to analyse cell structure and scaffold morphology.
- Atomic force microscopy to analyse the topography and elasticity of the scaffold.
- Micro-computed tomography (micro-CT) to analyse the three-dimensional structure of the cell-scaffold construct. It should be noted that the images obtained from micro-CT can be used in computational fluid dynamic programmes for more accurate modelling of the fluid dynamics.

The physical, chemical and biological data need to be related to fluid dynamics in a mathematical model so that spatial and temporal variations can be understood. Bioreactor measurements should first be taken without cells as a negative control, then with cells. Interactions of all chemical components with the scaffold should be measured. Fouling and loss of expensive components to the scaffold are important factors to quantify. Analysis to establish whether these reach equilibrium so the true concentrations of the drugs/toxins/growth factors/oxygen in the media, and at the cell surface, is vital.

3.6 Tissue engineering process design

Successful tissue engineering means that tissue manufacture moves from bench to factory, in a way similar to how penicillin moved to mass production during World War II (Neushul, 1993). As with any process design the final scale of production that is expected should be borne in mind from the outset, if you want to see your product in the clinic with minimal timeframe and maximal efficiency. For successful bioprocessing of cells a knowledge of how upstream and downstream processing are interlinked is necessary because choices made in the design of the upstream process affects the design of the downstream process. For example, knowing whether the end product is a vial of cells or a seeded scaffold will affect the choice of bioreactor and scaffold. A very good introduction to bioprocessing (albeit in protein purification) is given in Wheelwright (1991).

A systematic approach to the upstream design of a tissue-engineered product is shown in Fig. 3.7. The process should begin only when the desired

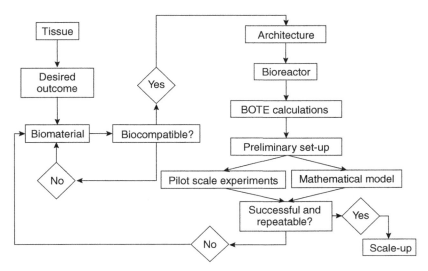

3.7 Bioreactor design flowchart. This diagram shows an approach to the bioprocess design steps for a bioreactor. 'BOTE' is the acronym for 'back of the envelope', rough calculations that can be done using mental arithmetic.

outcome is known, and the outcome should be known in detail, for example, a vial of 1 million cells, or a 2 cm diameter × 10 cm long biodegradable construct populated by the patient's own cardiomyocytes. In particular, the final scale should be at the forefront of your mind and while small-scale preliminary set-ups are space- and cost-saving it is important that the process of scale-up does not render the results invalid by conditions changing in such a way that the desired combination of conditions cannot be achieved. The tissue properties must be known (vascularised? ability to regenerate? load-bearing?), then the flowchart can be followed, using the tables for material, architecture and bioreactor selection above. Below is a hypothetical step-wise approach to engineering a liver regeneration construct.

1. It is known that 10% of the liver is required for survival, that is, 10 billion cells (Allen *et al.*, 2001; Sullivan *et al.*, 2007) and has the sinusoidal structure. So the end product should contain 10 billion cells and have good pseudovascularisation.
2. As liver is able to regenerate, choose a biodegradable material. It is known that PLGA is a suitable scaffold material (Kasuya *et al.*, 2012).
3. PLGA can be used to make a range of scaffold architectures. As liver is vascularised and non-loadbearing, a polymer will be suitable, and in an architecture that replicates vascularisation. Hollow fibre membranes in a hollow fibre bioreactor are a good starting set-up (Table 3.1).

48 Standardisation in cell and tissue engineering

4. Back of the envelope (BOTE) calculations are carried out to approximate the surface area needed and media requirements, based on standard T-flask culture.
5. Knowing that hollow fibres can be represented by Krogh cylinders (Brotherton and Chau, 1996; Shipley *et al.*, 2010) and are therefore relatively simple to scale up, a single-fibre module will be used as the preliminary set-up.
6. Experiments are run without cells to see what media components are fouling and lost to the fibres. Attachment and maintenance experiments are carried out to assess viability, mitochondrial metabolic activity and basic hepatocyte phenotype.
7. While attachment and maintenance experiments are carried out, the design is refined, using Shipley *et al.* (2011) to maintain $c > c_{min}$. Iteration is carried out to refine the experimental and mathematical model.
8. In-depth biological analysis is carried out.
9. Scale-up to several fibres, then the full-scale bioreactor is run and compared to the single-fibre model.
10. Once refined, the construct is ready for downstream processing.

3.7 Future trends

People are living longer, there is a rise in chronic diseases, and people are making lifestyle choices such as wishing to be more active later in life, so developments in tissue engineering are likely to play an important role in this rapidly changing healthcare landscape (BIS, 2011; MRC, 2012). Similarly, a demand for artificial assist organs to allow a stop-gap for a patient's own organ to recover or to permanently function will also increase. Tissue-engineered products have the potential for further use, as *in vitro* models to analyse tissue behaviour, for example in toxicology testing. Bioreactors can also be used for the expansion of cell populations that are harvested from the scaffold and supplied as the product. As the demand for tissue-engineered products increases, competition in the market increases and, as for any product, the most effective and most economic will prevail. Effectiveness relies on understanding the underlying science. Economic products rely on rigorous design methods to ensure there is no waste, that is, time, space, energy, labour and raw materials are used efficiently. Knowledge of the underlying science continues to grow, from understanding signalling pathways to the production of new biomaterials. Engineers are working with scientists to incorporate the new knowledge into modified bioreactor system design. Measurements of the bioreactor environment and cell behaviour within it are vital, and progress in the biosensor field has the potential to allow real-time data to be obtained for an array of biological parameters. As such, future trends will likely focus on the micro- and nano-scale

© Woodhead Publishing Limited, 2013

3.8 Conclusion

Tissue engineering is complex and requires an in-depth understanding of all components of the *in vivo* and *in vitro* environment. It can be seen from the level of detail required in engineering, chemistry and materials, maths, and biology, and also from the clinicians' point of view, that tissue engineering is highly interdisciplinary. The importance of working with experts in numerous other fields should be evident. A bioprocessing approach to bioreactor design for soft tissue engineering enables the tissue engineer to establish the design in a systematic manor, incorporating all the areas of expertise, which should ultimately result in a more efficient, economic and robust product.

3.9 Sources of further information and advice

Sources of information on bioreactor design are, fortunately, numerous. There is less literature specifically on bioreactor design for tissue engineering; therefore the available literature must be used in conjunction with other sources. Standard biochemical engineering texts that provide an excellent introduction are Bailey and Ollis (1986) and Doran (1997). The comprehensive Coulson and Richardson's Chemical Engineering series, particularly Book 1, 2 and 6 (Coulson *et al.*, 1996, 1997; Sinnott, 2005) provides an excellent foundation to the field. Scragg (1991) covers general bioreactor design in more detail. Cussler (1984) is a good text to introduce mass transfer and fluid dynamics, and for a more in-depth text the reader is direct toward Kay and Nedderman (1974), and for tissue engineering, Saltzman (2004); Lanza *et al.* (2007). Some newer texts contain useful information on bioreactor design for tissue engineering including Chaudhuri and Al-Rubeai (2005) and Ellis *et al.* (2008). There are also many journals that publish tissue engineering-related articles and the reader should be willing to look at these and less obvious sources that are specific to certain aspects of design; for example *Journal of Membrane Science*, currently the leading membrane journal, has some very useful papers that are relevant for membrane bioreactor design, *Biotechnology and Bioengineering* covers the broader aspects of biochemical engineering including tissue engineering with a more engineering and mathematical slant. A similar approach should be taken when attending conferences; look to attend those whose focus is on specific aspects of the design as well as broad tissue engineering meetings.

Organisations under which bioreactor design is housed are the Institution of Chemical Engineers (IChemE) and their Biochemical Engineering Special

50 Standardisation in cell and tissue engineering

Interest Group (BESIG), in Africa, Asia and the Middle East, Australasia and Europe; the American Institution of Chemical Engineers (AIChemE) in North America and 93 countries in total; and the European Federation of Biotechnology (EFB) European Section on Biochemical Engineering Science (ESBES) in Europe.

3.10 References

Allen, J. W. and Bhatia, S. N. 2003. Formation of steady-state oxygen gradients *in vitro*: application to liver zonation. *Biotechnology and Bioengineering*, **82**, 253–262.

Allen, J. W., Hassanein, T. and Bhatia, S. N. 2001. Advances in bioartificial liver devices. *Hepatology*, **34**, 447–455.

Badylak, S. F. 2004. Xenogeneic extracellular matrix as a scaffold for tissue reconstruction. *Transplant Immunology*, **12**, 367–377.

Bailey, J. E. and Ollis, D. F. 1986. *Biochemical Engineering Fundamentals*, Singapore, McGraw-Hill Inc.

Bambrick, L. L., Kostov, Y. and Rao, G. 2011. *In vitro* cell culture pO_2 is significantly different from incubator pO_2. *Biotechnology Progress*, **27**, 1185–1189.

Berridge, M. J. 2009. *Cell Signalling Biology* (Online). Available: http://www.biochemj.org/csb/ (Accessed 22 March 2012).

Biggs, M. J. P., Richards, R. G. and Dalby, M. J. 2010. Nanotopographical modification: a regulator of cellular function through focal adhesions. *Nanomedicine: Nanotechnology, Biology, and Medicine*, **6**, 619–633.

BIS. 2011. *Strategy for UK Life Sciences* (Online). Department for Business, Innovation and Skills. Available: http://www.bis.gov.uk/assets/biscore/innovation/docs/s/11-1429-strategy-for-uk-life-sciences.pdf.

Bray, L. J., George, K. A., Hutmacher, D. W., Chirila, T. V. and Harkin, D. G. 2012. A dual-layer silk fibroin scaffold for reconstructing the human corneal limbus. *Biomaterials*, **33**, 3529–3538.

Brotherton, J. D. and Chau, P. C. 1996. Review: modeling of axial-flow hollow fiber cell culture bioreactors. *Biotechnology Progress*, **12**, 575–590.

Chaudhuri, J. and Al-Rubeai, M. 2005. *Bioreactors for Tissue Engineering: Principles, Design and Operation*, Dordrecht, Springer.

Coulson, J. M., Richardson, J. F., Bankhiurst, J. R. and Harker, J. H. 1996. *Coulson and Richardson's Chemical Engineering: Particle Technology and Separation Processes*, Oxford, Butterworth-Heinemann Ltd.

Coulson, J. M., Richardson, J. F., Backhurst, J. R. and Harker, J. 1997. *Coulson and Richardson's Chemical Engineering: Fluid Flow, Heat Transfer and Mass Transfer*, Oxford, Butterworth-Heinemann Ltd.

Curcio, E., De Bartolo, L., Barbieri, G., Rende, M., Giorno, L., Morelli, S. and Drioli, E. 2005. Diffusive and convective transport through hollow fiber membranes for liver cell culture. *Journal of Biotechnology*, **117**, 309–321.

Cussler, E. L. 1984. *Diffusion: Mass Transfer in Fluid Systems*, Cambridge, Cambridge University Press.

Dalby, M. J., Childs, S., Riehle, M. O., Johnstone, H. J. H., Affrossman, S. and Curtis, A. S. G. 2003. Fibroblast reaction to island topography: changes in cytoskeleton and morphology with time. *Biomaterials*, **24**, 927–935.

Davidson, A. J., Ellis, M. J. and Chaudhuri, J. B. 2012. A theoretical approach to zonation in a bioartificial liver. *Biotechnology and Bioengineering*, **109**, 234–243.

Discher, D. E., Janmey, P. and Wang, Y. L. 2005. Tissue cells feel and respond to the stiffness of their substrate. *Science*, **310**, 1139–1143.

Doran, P. M. 1997. *Bioprocess Engineering Principles*, London, Academic Press Limited.

Ellis, M. J. and Chaudhuri, J. B. 2007. Poly(lactic-co-glycolic acid) hollow fibre membranes for use as a tissue engineering scaffold. *Biotechnology and Bioengineering*, **96**, 177–187.

Ellis, M. J., De Bank, P. A. and Jones, M. D. 2008. Polymeric Scaffolds in Regenerative Medicine. In: Kumar, A. (ed.) *Macroporous Polymers: Production, Properties and Biotechnological/Biomedical Applications*, 1 ed. Boca Raton, Taylor & Francis Group.

Fagerholm, P., Lagali, N. S., Merrett, K., Jackson, W. B., Munger, R., Liu, Y., Polarek, J. W., Söderqvist, M. and Griffith, M. 2010. A biosynthetic alternative to human donor tissue for inducing corneal regeneration: 24-Month follow-up of a phase 1 clinical study. *Science Translational Medicine*, **2**, 46ra61.

Grek, C. L., Newton, D. A., Qiu, Y., Wen, X., Spyropoulos, D. D. and Baatz, J. E. 2009. Characterization of alveolar epithelial cells cultured in semipermeable hollow fibers. *Experimental Lung Research*, **35**, 155–174.

Jones, G. L., Walton, R., Czernuszka, J., Griffiths, S. L., El Haj, A. J. and Cartmell, S. H. 2010. Primary human osteoblast culture on 3D porous collagen-hydroxyapatite scaffolds. *Journal of Biomedical Materials Research – Part A*, **94**, 1244–1250.

Jones, R. R., Hamley, I. W. and Connon, C. J. 2012. *Ex vivo* expansion of limbal stem cells is affected by substrate properties. *Stem Cell Research*, **8**, 403–409.

Kasuya, J., Sudo, R., Tamogami, R., Masuda, G., Mitaka, T., Ikeda, M. and Tanishita, K. 2012. Reconstruction of 3D stacked hepatocyte tissues using degradable, microporous poly(d,l-lactide-co-glycolide) membranes. *Biomaterials*, **33**, 2693–2700.

Kay, J. M. and Nedderman, R. M. 1974. *An Introduction to Fluid Mechanics and Heat Transfer: With Applications in Chemical & Mechanical Process Engineering*, Cambridge, Cambridge University Press.

Kimura, Y., Ozeki, M., Inamoto, T. and Tabata, Y. 2003. Adipose tissue engineering based on human preadipocytes combined with gelatin microspheres containing basic fibroblast growth factor. *Biomaterials*, **24**, 2513–2521.

Lanza, R., Langer, R. and Vacanti, J. 2007. *Principles of Tissue Engineering*, Burlington, MA, Elsevier Academic Press.

Lawrence, B. D., Marchant, J. K., Pindrus, M. A., Omenetto, F. G. and Kaplan, D. L. 2009. Silk film biomaterials for cornea tissue engineering. *Biomaterials*, **30**, 1299–1308.

Lim, J. I. and Park, H. K. 2011. Fabrication of macroporous chitosan/poly(l-lactide) hybrid scaffolds by sodium acetate particulate-leaching method. *Journal of Porous Materials*, **19**(3), 383–387.

Liu, Z., Wang, H., Wang, Y., Lin, Q., Yao, A., Cao, F., Li, D., Zhou, J., Duan, C., Du, Z., Wang, Y. and Wang, C. 2012. The influence of chitosan hydrogel on stem cell engraftment, survival and homing in the ischemic myocardial microenvironment. *Biomaterials*, **33**, 3093–3106.

Lu, H. F., Lim, W. S., Zhang, P. C., Chia, S. M., Yu, H., Mao, H. Q. and Leong, K. W. 2005. Galactosylated poly(vinylidene difluoride) hollow fiber bioreactor for hepatocyte culture. *Tissue Engineering*, **11**, 1667–1677.

52 Standardisation in cell and tissue engineering

Macaya, D. and Spector, M. 2012. Injectable hydrogel materials for spinal cord regeneration: A review. *Biomedical Materials*, **7**, 012001.

McNamara, L. E., Burchmore, R., Riehle, M. O., Herzyk, P., Biggs, M. J. P., Wilkinson, C. D. W., Curtis, A. S. G. and Dalby, M. J. 2012. The role of microtopography in cellular mechanotransduction. *Biomaterials*, **33**, 2835–2847.

Morgan, S. M., Ainsworth, B. J., Kanczler, J. M., Babister, J. C., Chaudhuri, J. B. and Oreffo, R. O. C. 2009. Formation of a human-derived fat tissue layer in PdlLGA hollow fibre scaffolds for adipocyte tissue engineering. *Biomaterials*, **30**, 1910–1917.

MRC. 2012. *A Strategy for UK Regenerative Medicine* (Online). Medical Research Council. Available: http://www.mrc.ac.uk/Utilities/Documentrecord/index.htm?d=MRC008534 (Accessed 29 March 2012).

Murphy, C. M., Matsiko, A., Haugh, M. G., Gleeson, J. P. and O'brien, F. J. 2011. Mesenchymal stem cell fate is regulated by the composition and mechanical properties of collagen-glycosaminoglycan scaffolds. *Journal of the Mechanical Behavior of Biomedical Materials*, **11**(Special Issue), 53–62.

Nair, L. S. and Laurencin, C. T. 2006. Polymers as biomaterials for tissue engineering and controlled drug delivery. In: Lee, K. and Kaplan, D. (eds) *Tissue Engineering I*. Berlin, Heidelberg, Springer-Verlag.

Neushul, P. 1993. Science, government and the mass production of penicillin. *Journal of the History of Medicine and Allied Sciences*, **48**, 371–395.

Newman, K. D. and McBurney, M. W. 2004. Poly(D,L lactic-co-glycolic acid) microspheres as biodegradable microcarriers for pluripotent stem cells. *Biomaterials*, **25**, 5763–5771.

Oo, Z. Y., Deng, R., Hu, M., Ni, M., Kandasamy, K., Bin Ibrahim, M. S., Ying, J. Y. and Zink, D. 2011. The performance of primary human renal cells in hollow fiber bioreactors for bioartificial kidneys. *Biomaterials*, **32**, 8806–8815.

Oudega, M., Gautier, S. E., Chapon, P., Fragoso, M., Bates, M. L., Parel, J. M. and Bartlett Bunge, M. 2001. Axonal regeneration into Schwann cell grafts within resorbable poly(alpha-hydroxyacid) guidance channels in the adult rat spinal cord. *Biomaterials*, **22**, 1125–1136.

Pek, Y. S., Wan, A. C. A. and Ying, J. Y. 2010. The effect of matrix stiffness on mesenchymal stem cell differentiation in a 3D thixotropic gel. *Biomaterials*, **31**, 385–391.

Perlin, L., Macneil, S. and Rimmer, S. 2008. Production and performance of biomaterials containing RGD peptides. *Soft Matter*, **4**, 2331–2349.

Redick, S. D., Settles, D. L., Briscoe, G. and Erickson, H. P. 2000. Defining Fibronectin's cell adhesion synergy site by site-directed mutagenesis. *Journal of Cell Biology*, **149**, 521–527.

Salerno, A., Di Maio, E., Iannace, S. and Netti, P. A. 2011. Tailoring the pore structure of PCL scaffolds for tissue engineering prepared via gas foaming of multi-phase blends. *Journal of Porous Materials*, **19**(2), 181–188.

Saltzman, W. M. 2001. *Drug Delivery: Engineering Principles for Drug Therapy*. New York, Oxford University Press.

Saltzman, W. M. 2004. *Tissue Engineering: Engineering Principles for the Design of Replacement Organs and Tissues*. New York, Oxford University Press.

Scragg, A. H. (ed.) 1991. *Bioreactors in Biotechnology: A Practical Approach*. Chichester, West Sussex, Ellis Horwood Limited.

Shipley, R. J., Davidson, A. J., Chan, K., Chaudhuri, J. B., Waters, S. L. and Ellis, M. J. 2011. A strategy to determine operating parameters in tissue engineering hollow fiber bioreactors. *Biotechnology and Bioengineering*, **108**, 1450–1461.

© Woodhead Publishing Limited, 2013

Shipley, R. J., Waters, S. L. and Ellis, M. J. 2010. Definition and validation of operating equations for poly(vinyl alcohol)-poly(lactide-co-glycolide) microfiltration membrane-scaffold bioreactors. *Biotechnology and Bioengineering*, **107**, 382–392.

Sinnott, R. K. 2005. *Coulson and Richardson's Chemical Engineering. Volume 6, Chemical Engineering Design*. Oxford, Butterworth-Heinemann.

Soffer, L., Wang, X., Zhang, X., Kluge, J., Dorfmann, L., Kaplan, D. L. and Leisk, G. 2008. Silk-based electrospun tubular scaffolds for tissue-engineered vascular grafts. *Journal of Biomaterials Science, Polymer Edition*, **19**, 653–664.

Storm, M. P., Orchard, C. B., Bone, H. K., Chaudhuri, J. B. and Welham, M. J. 2010. Three-dimensional culture systems for the expansion of pluripotent embryonic stem cells. *Biotechnology and Bioengineering*, **107**, 683–695.

Sullivan, J. P., Gordon, J. E., Bou-Akl, T., Matthew, H. W. T. and Palmer, A. F. 2007. Enhanced oxygen delivery to primary hepatocytes within a hollow fiber bioreactor facilitated via hemoglobin-based oxygen carriers. *Artificial Cells, Blood Substitutes, and Biotechnology*, **35**, 585–606.

Varki, A. 1997. Sialic acids as ligands in recognition phenomena. *FASEB Journal*, **11**, 248–255.

Wheelwright, S. M. 1991. *Protein Purification: Design and Scale Up of Downstream Processing*. Hanser, Muncih: Wiley.

Wion, D., Christen, T., Barbier, E. L. and Coles, J. A. 2009. PO_2 matters in stem cell culture. *Cell Stem Cell*, **5**, 242–243.

Wu, I., Nahas, Z., Kimmerling, K. A., Rosson, G. D. and Elisseeff, J. H. 2012. An injectable adipose matrix for soft tissue reconstruction. *Plastic and Reconstructive Surgery*, **129**, 1247–1257.

Wu, X., Jones, M. D., Davidson, M. G., Chaudhuri, J. B. and Ellis, M. J. 2010. Surfactant-free poly(lactide-co-glycolide) honeycomb films for tissue engineering: relating solvent, monomer ratio and humidity to scaffold structure. *Biotechnology Letters*, accepted.

Yang, F., Murugan, R., Wang, S. and Ramakrishna, S. 2005. Electrospinning of nano/micro scale poly(l-lactic acid) aligned fibers and their potential in neural tissue engineering. *Biomaterials*, **26**, 2603–2610.

Yang, X. B., Roach, H. I., Clarke, N. M. P., Howdle, S. M., Quirk, R., Shakesheff, K. M. and Oreffo, R. O. C. 2001. Human osteoprogenitor growth and differentiation on synthetic biodegradable structures after surface modification. *Bone*, **29**, 523–531.

Yu, S. C., Park, S. N. and Suh, H. 2005. Adipose tissue engineering using mesenchymal stem cells attached to injectable PLGA spheres. *Biomaterials*, **26**, 5855–5863.

Zhu, N., Warner, R. M., Simpson, C., Glover, M., Hernon, C. A., Kelly, J., Fraser, S., Brotherston, T. M., Ralston, D. R. and Macneil, S. 2005. Treatment of burns and chronic wounds using a new cell transfer dressing for delivery of autologous keratinocytes. *European Journal of Plastic Surgery*, **28**, 319–330.

Zhu, X., Cui, W., Li, X. and Jin, Y. 2008. Electrospun fibrous mats with high porosity as potential scaffolds for skin tissue engineering. *Biomacromolecules*, **9**, 1795–1801.

4

Two- and three-dimensional tissue culture methods for hard tissue engineering

M. A. BIRCH and K. E. WRIGHT, Newcastle University, UK

DOI: 10.1533/9780857098726.1.54

Abstract: A complex array of cellular and molecular events underlies the (patho)physiology of bone and cartilage. Tissue culture approaches have provided great insight into the mechanisms that regulate the activity of bone cells (osteoclasts, osteoblasts, osteocytes) and cartilage cells (chondrocytes). These approaches include assays that have evolved to investigate the differentiation processes of progenitor cells as well as characterising the functionality of mature cells. What has become increasingly apparent is that the three-dimensional environment is an important regulator in the control of cell activity, and tissue culture approaches have been developed to evaluate the role of these substrate-derived cues.

Key words: osteoblast, osteoclast, osteocyte, chondrocyte, mesenchymal stem cell.

4.1 Introduction

Bone and cartilage are specialised connective tissues that combine extracellular matrix (ECM) biomolecules and complexes of inorganic ions to give mechanically competent and structurally durable materials that meet functional requirements. The formation, development and maintenance of these tissues are dependent on the coordinated activity of several different cell types. Much of our knowledge and understanding of the cell biology that underpins these events has been derived from experimental approaches that utilise cell culture as a key tool to investigate the underlying molecular mechanisms. In addition, the possibility of using *in vitro* expanded cells in tissue engineering treatment strategies to repair diseased or damaged tissues has led to the further development and refinement of culture conditions. In this chapter we explore the techniques and methodology that are employed for the growth and study of those cells responsible for hard tissue biology.

54

© Woodhead Publishing Limited, 2013

4.1.1 Background

Bone is structured at the macro, micro and nano scale to meet its functional roles. The macro structure of bone is characterised by a tough durable outer layer that is termed cortical (compact) bone whilst the inner marrow cavity is criss-crossed and reinforced by struts of trabecular (cancellous) bone. At the microscale the structural unit in compact bone is the haversian system where cylinders of mineralised matrix surround a central canal that contains a blood vessel, whilst at the nanoscale collagen fibrils are arranged in parallel bundles onto which hydroxyapapatite $(Ca_{10}(PO_4)_6(OH)_2)$ is deposited to make bone a mechanically strong biocomposite material.

Cells of bone exhibit a regional distribution that underlies their role in regulating bone mass. Osteocytes are entombed within tiny lacunae in compact bone; they remain in contact with each other and with cells at the surface through cellular processes that extend through canaliculi. At the surface of bones, cells from the osteoblast lineage are in a resting state and appear flattened and more rounded. These bone surface osteoblasts are intensely alkaline phosphatase positive and actively synthesising mineralised matrix. In addition, at discrete foci at the bone surface, cells of haematopoietic origin may be found. These haematopoietic cells fuse to form multinucleated osteoclasts that are capable of breaking down and resorbing all of the components of bone. The activities of these cells are controlled and coordinated by hormones, cytokines and growth factors, together with immobilised signals from the matrix components.

Cartilage is found principally at the junctions and articulations between bones, helping to provide smooth movement of the joints and resistance to load. In addition, cartilage forms the nasal septum, larynx, trachea and bronchi of the respiratory tract. Cartilage is principally composed of fibrillar type II collagen and the large aggregating proteoglycan, aggrecan. Collagen is arranged to provide great tensile strength while the highly charged glycosaminoglycan side chains of aggrecan attract water into the tissue, ensuring its resistance to compressive load. In cartilage there is a single cell type, the chondrocyte, which can be found embedded within a pericellular matrix secreted from surrounded cells. Cartilage is an avascular tissue so chondrocytes are dependent on nutrient diffusion from synovial fluid and surrounding tissues.

4.1.2 Tissue culture modelling of bone and cartilage *in vitro*

Bone and cartilage cell activity has been investigated using *ex vivo* organ cultures, *in vitro* primary and/or immortalised cells, and cells in combination with supporting scaffolds to mimic the *in vivo* environment. All these

56 Standardisation in cell and tissue engineering

approaches have been further refined with the development of techniques to deliver mechanical load, hydrostatic pressure and shear force to recapitulate some elements of the physical environment found *in vivo* (McGlashan *et al.*, 2010; Kumbar *et al.*, 2011).

Ex vivo organ cultures are explants of tissue, for example, bone, cartilage, etc. that can be used to investigate aspects of cell biology. Isolation of tissue from its vascular supply means that nutrients and waste diffusion can become an issue within the explanted tissue which is dependent on its size. In addition the limited supply of haematopoietic cells means that events like osteoclastogenesis cannot really be investigated. Assay models that have classically been investigated include the use of calvariae cultures to assess osteoclast and osteoblast function, as well as cartilage plugs to investigate chondrocyte activity.

Immortalised cell lines have largely been derived from tumour material and have been mostly used for modelling aspects of osteoblast and chondrocyte activity. Osteosarcoma cell lines have been shown to appropriately respond to parathyroid hormone (PTH), $1,25(OH)_2$ vitamin D_3 and other bone active factors, but do not recapitulate all of the events that lead to the formation of a mineralised bone-like matrix *in vitro*. Similarly, chondrosarcoma cell lines are proliferative and, while used to assess aspects of chondrocyte function, clearly represent a partially dedifferentiated phenotype. Reports have suggested that some cell lines of haematopoietic origin can be induced to form osteoclasts. Murine monocytes (RAW 264.7 cells) have been the most widely used precursors to form functional osteoclasts *in vitro*. Cell lines have also been developed by the introduction of Simian Virus-40 large T antigen or human telomerase reverse transcriptase (hTERT). Using these approaches, cell lines have been created that allow the study of osteocytes, osteoblasts and chondrocytes without neoplastic transformation.

Isolated primary cells have long been used *in vitro* to investigate the cell biology of bone and cartilage. Mature populations of cells such as osteoclasts, osteoblasts and chondrocytes can be harvested from tissue and their activity directly investigated *in vitro*. Alternatively populations of progenitor cells can be extracted from primary tissue samples and used to study the mechanisms associated with the differentiation process, for example, osteogenesis and osteoclastogenesis.

4.2 Culture of bone and cartilage cells

Establishing primary cultures from musculoskeletal tissue is one approach to the study of bone and cartilage cells.

4.2.1 Tissue sources for cell harvesting

Human or other animal skeletal tissue can be used as sources for bone and cartilage cells. Donor age and harvest site both contribute to the ease of

2- and 3-D tissue culture methods for hard tissue engineering   57

processing, cell isolation and ultimate suitability of these cells for an experimental approach. For example, some protocols may yield mixed populations of cells, whilst others may depend upon significant culture expansion with the caveat of phenotypic drift. Using embryonic or neonatal tissues can ease the process of isolation, since the softer bones can be more readily diced into manageable fragments. Furthermore, cells from younger donors tend to exhibit a longer lifespan in culture and a more rapid doubling rate. Populations of cells may simply be retrieved by mechanical agitation of the tissue fragments in a suitable media preparation or may be liberated through multiple rounds of enzymatic and/or nonenzymatic tissue digests.

Whilst mature cells, committed progenitors and some stem cells can be isolated directly from musculoskeletal tissue, mesenchymal and haematopoietic stem cells may also be derived from other tissues for bone and cartilage research. These include, but are not limited to peripheral blood, a source of mononuclear cells that can be induced using appropriate culture conditions to form osteoclasts; from bone marrow mesenchymal stem cells (MSCs) which exhibit multipotential differentiation, including differentiation into chondrocytes and cells of the osteoblast lineage; synovium, dental pulp and adipose tissue; and post-natal sources (umbilical cord, placenta, amniotic fluid and cord blood).

4.2.2 Cell isolation

Primary cell isolation usually requires the dissection of tissue from a primary tissue source. The exception to this is the use of blood or bone marrow for the isolation of haematopoietic or other mononuclear cell populations such as MSCs. Explanted tissue is then subjected to nonenzymatic (mechanical) and/or enzymatic treatment in order to release cells into a suspension. A common routine approach would be to first dice the tissue into small fragments and then expose them to an enzyme or enzyme cocktail such as collagenase, hyaluronidase or pronase, at 37°C with agitation for a period of time (Sauren *et al.*, 1989; Semeins *et al.*, 2012). Alternatively, small pieces of explanted tissue can be cultured to allow cell outgrowth or migration from the tissue (Chan *et al.*, 2009). Isolated cells are allowed to proliferate or expand *in vitro* with the use of a specially formulated expansion medium.

4.2.3 Media formulations

The choice of cell culture media is largely empirical and often reflects local laboratory experiences. As a result there are a large number of in-house media formulations as well as commercially available media for a given application. In-house recipes are commonly based on media such as alpha Minimal Essential Medium (α-MEM), Dulbecco's Modified Eagles Medium

58 Standardisation in cell and tissue engineering

(DMEM) and Roswell Park Memorial Institute medium (RPMI). For cell growth and expansion these basal media are generally supplemented with animal serum, a source of glutamine, antibiotics and sometimes antimycotics. Varying types and concentrations of growth factors have also been added to these basal media in order to maintain cell viability, proliferation, phenotype, differentiation states and ECM deposition.

Commercially formulated media sources tend to provide a tailored solution for the growth and expansion of a particular cell type and are often targeted to workers in the fields of tissue engineering and regenerative medicine. These media are marketed as being optimised for the isolation, expansion and differentiation of a given cell population. However, the components of such media formulations may be unavailable to the user. This makes these sources of media problematic options for use in some experiments, that is, experiments where the exact components of the culture media are required to be disclosed to comply with good manufacturing practices and for the assessment of suitability for clinical implementation of cells and tissue-engineered constructs.

4.2.4 Cell characterisation

Bone cells

Osteocytes can be characterised by the detection of PHEX with the antibody MAB OB 7.3 for osteocytic cells (Westbroek *et al.*, 2002). They also display positivity for DMP1, MEPE and scelerostin; negativity for osteoblast-specific factor 2 (OSF-2); and have high amounts of osteocalcin with low amounts of alkaline phosphatase (Kato *et al.*, 1997; Javed *et al.*, 2010).

Osteoblasts have a cuboidal shape (Bakker and Klein-Nulend, 2003) particularly *in vivo*, but *in vitro* cell morphology correlates poorly with osteoblast phenotype. Osteocalcin (bone-gla protein) is the most specific marker associated with osteoblasts and is usually supported with alkaline phosphatase positivity (Gallagher *et al.*, 1996; Gartland *et al.*, 2012). After long-term culture (2–4 weeks) the formation of nodules (see Plate I in colour section between pages 134 and 135) that are positive for minerals with the use of Von Kossa or Alizarin Red S staining are an excellent confirmation of differentiation to mature osteoblast. Osteoblasts also display a characteristic increase in intracellular cyclic adenosine monophosphate (cAMP) after stimulation with PTH stimulation (Hillsley, 1999). Commitment to the osteoblast lineage and subsequent differentiation is also often characterised in terms of transcript levels for transcription factors, such as Runx2 and osterix as well as ECM proteins like osteocalcin, type I collagen and osteopontin. Osteoprogenitor cells have been identified *in vivo* using the activity of the alpha smooth muscle actin promoter (San Miguel *et al.*, 2009). Positivity for

© Woodhead Publishing Limited, 2013

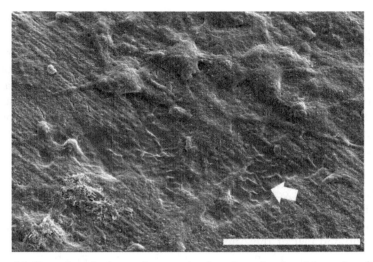

4.1 Scanning electron micrograph of a mineralised matrix wafer after use as an osteoclast cell culture substrate. Resorption lacunae (white arrow) are clearly visible as erosions that contrast the machining lines of the cut surface (scale bar = 50 microns).

ALCAM and SDF-1 are also indicators of osteoprogenitors (Javed *et al.*, 2010).

Multinuclearity clearly provides some evidence of the osteoclast phenotype *in vitro* but does not provide absolute certainty. Commonly used markers to identify osteoclasts include tartrate-resistant acid phosphatase (see Plate II in colour section between pages 134 and 135), the vitronectin receptor CD51, calcitonin receptor and the inhibition of osteoclast resorption using calcitonin (Sabokbar and Athanasou, 2003). Of course unequivocal evidence for osteoclasts is the characteristic resorption of a devitalised mineralised matrix such as dentine. Resorption lacunae can be visualised by toluidine blue staining, reflected light microscopy or by scanning electron microscopy (Fig. 4.1).

Cartilage cells

Chondrogenic micromass can be investigated after 10% (w/v) formalin fixation, paraffin embedding and histological analysis. Micromass positivity for cartilage development is assessed on the basis of multilayered matrix-rich morphology and histology. An increase in the level of proteoglycan-rich ECM and chondrocyte-like lacunae can be observed in histological sections (Mackay *et al.*, 1998). The ECM should be rich in aggrecan and type II collagen that are found in articular cartilage and detectable at day 10 and 14 of culture (Mackay *et al.*, 1998). Thin sections (4–10 μm) of micromass

60   Standardisation in cell and tissue engineering

can be analysed by staining with Alcian Blue and Safranin O which both localise complex polysaccharides and glycosaminoglycans such as aggrecan. Additional evidence for the chondrocytic phenotype can be provided by analysis of mRNA transcript profiles. Expression of type II collagen, Sox9, aggrecan and type X collagen (hypertrophic chondrocytes) has been reported as representative of the chondrocytic phenotype, while absence of type I collagen has been useful in illustrating the lack of phenotypic drift within this cell population.

4.3   Cell culture parameters: bone tissue culture

All types of bone cell can be studied *in vitro* including osteoprogenitors (committed yet immature cells which can go on to differentiate into osteoblasts), osteoblasts (mature bone-forming cells), osteocytes (a terminally differentiated cell of the osteoblast lineage) and osteoclasts (responsible for the resorption of bone). Described below are commonly used cell lines and primary cell cultures used to understand the biology of bone as well as for the basis for tissue engineering approaches.

4.3.1 Cell lines

Commonly used and readily available bone cell lines to model osteoblast and osteocyte activity include: human osteosarcoma cell lines (MG-63, SaOS-2, and TE-85), murine osteoblastic cells lines (MC3T3 and MCT3T3-E1) and a mouse osteocyte cell line (MLO-Y4). All of these cell lines can be cultured in routine culture media with a general composition: α-MEM supplemented with 5–10% (v/v) fetal bovine serum (FBS), 2 mM L-glutamine, 100 IU/mL penicillin, 100 mg/mL streptomycin, 2.5 mg/mL amphotericin B, and 50 mg/mL ascorbic acid under cell culture conditions at 37°C in a humidified 5% CO_2 atmosphere (Bliziotes *et al.*, 2006; Costa-Rodrigues *et al.*, 2011; Han *et al.*, 2011). These cell lines can typically be seeded at $1 \times 10^2 - 1 \times 10^4$ cells/mL and are plastic-adherent and therefore require trypsinisation to allow for sub-culturing.

RAW 264.7 is a macrophage-like cell line that under appropriate culture conditions has been demonstrated to differentiate into osteoclast-like cells *in vitro* (Wei *et al.*, 2001). Routine culture of this cell line is in 1:1 DMEM:Ham's F-12 medium mixture or DMEM supplemented with 2–4 mM L-glutamine and 10% (v/v) FBS (Collin-Osdoby and Osdoby, 2012). Initial seeding density is of the order 5×10^5 cells/mL. Since adhesion to plastic ware is a strong signal for these cells to differentiate, routine culturing of these cells should be performed in non-tissue culture grade plastic. Differentiation of these cells to the osteoclast lineage can be achieved by supplementing of the culture medium with 35–50 ng/mL RANK (receptor activator of NF-κB)

© Woodhead Publishing Limited, 2013

2- and 3-D tissue culture methods for hard tissue engineering 61

ligand (Collin-Osdoby and Osdoby, 2012). After approximately one week multinucleated cells that exhibit many properties of osteoclasts become visible in the culture system.

4.3.2 Primary cultures from isolated tissues

Osteocytes and osteoblasts

Osteocytes are embedded in the bone matrix and are the most abundant cells in bone (Nijweide *et al.*, 2003; Chan *et al.*, 2009). Osteocytes represent approximately 90% of all bone cells (Bonewald, 1999). Sequential enzymatic digests at 37°C with 1 mg/mL collagenase and 4 mM EDTA of embryonic chick calvaria results in the isolation of osteocytes in conjunction with some osteoblasts (Semeins *et al.*, 2012). Osteocytes can be separated from the heterogenous cell suspension by using the osteocyte-specific antibody MAb OB7.3 (detects PHEX) in an immunomagnetic separation system (Westbroek *et al.*, 2002; Semeins *et al.*, 2012). Primary osteocytes are cultured in α-MEM, 2% (v/v) serum, 0.2 g/L glutamine, 0.05 g/L ascorbic acid, 0.05 g/L gentamycin and 1 g/L glucose (Semeins *et al.*, 2012).

Primary osteoblasts can be cultured in commercially available medium such as OGM™ Osteoblast Growth Media BulletKit™ (Lonza) but their growth is usually well supported by routine culture media based on DMEM or α-MEM. These cells can be isolated or outgrown from several sources including trabecular bone chips and bone cores. Trabecular bone chips can be washed free of adherent bone marrow cells by agitation in phosphate-buffered saline (PBS) and cultured first for 7 days (to remove unwanted heterogenous cell populations) in commercially available medium or in-house generated culture medium consisting of α-MEM, 100 U/mL penicillin, 100 µg/mL streptomycin and 10% (v/v) FBS. The small bone chips can then be transferred to a new culture dish and cultured for a further 2–3 weeks and this second population of cells are enriched for osteoblastic cells (Chan *et al.*, 2009). An alternative source of osteoblasts is from bone cores which are harvested and rinsed in PBS, then seeded and cultured statically for 24 h to allow for primary osteoblast attachment. These cells are cultured in osteogenic medium, which is routine culture media supplemented with 10 or 100 nmol/L dexamethasone, 50 µg/mL or 0.2 mmol/L ascorbic acid and 10 mmol/L β-glycerophosphate, before transferring to a medium perfusion system (Chan *et al.*, 2009; Yu *et al.*, 2009; Gartland *et al.*, 2012).

Osteoprogenitors

Calvariae from 2- to 3-days-old Sprague-Dawley rats are harvested by dissection and diced into small fragments (Yamamoto *et al.*, 2002). Calvariae

62   Standardisation in cell and tissue engineering

contain a range of undifferentiated mesenchymal osteoprogenitor cells through to PTH-responsive osteoblasts. Calvaria fragments are digested with five sequential 20 min digests (I to V) at 37°C in Hefley's buffer containing 2 mg/mL collagenase B and 0.25% (w/v) Trypsin (Hefley *et al.*, 1981; Yamamoto *et al.*, 2002; Owen and Pan, 2008). Single cell suspensions are pooled from digests III to V and plated on day zero in MEM supplemented with 10% (v/v) FBS and antibiotics (100 U/mL penicillin G and 100 µg/mL streptomycin) (Yamamoto *et al.*, 2002). These populations of cells can be expanded and if used at early passage number display multipotency to lineages that include bone, cartilage and fat.

These osteoprogenitor cells can be differentiated into mature osteoblasts with the use of a differentiation medium such as α-MEM supplemented with 10% (v/v) FBS, 100 µg/ mL ascorbic acid, 0.5 M sodium phosphate and 10^{-9} M dexamethasone (Owen and Pan, 2008). There are some exceptions such as murine osteoprogenitor cells where dexamethasone seems not to be required for differentiation to osteoblasts (Owen and Pan, 2008).

Osteoclasts

Osteoclasts are large multinucleated cells that are normally found in small numbers opposed to the surface of bone. As a consequence isolation of mature cells, particularly of human origin, can be problematic. Human osteoclasts can be retrieved in sufficient numbers for experimentation from osteoclastoma tissue but this benign tumour is rare and therefore impractical for routine laboratory experimentation. Embryonic or neonatal bones from several species (chick, mouse or rat) can be isolated and the marrow contents flushed from long bones where the epiphyses have been removed. This mixed marrow suspension that includes mature osteoclasts can be cultured in α-MEM-based media in experimental approaches that address osteoclast activity, for example, culture on devitalised mineralised matrix.

Osteoclasts are formed from mononucleated cells of haematopoietic origin and the process of osteoclastogenesis can be studied by the culture of these cells. Density gradient centrifugation can be used to separate out mononuclear cells from human peripheral blood (Susa *et al.*, 2004; Agrawal *et al.*, 2011). The mononuclear cells are then cultured for 17–21 days at a seeding density of approximately 600 000 cells per well in a 96-well plate or on dentine discs in α-MEM medium, supplemented with 10% (v/v) FBS, 1 µM dexamethasone, and 25 ng/mL macrophage-colony stimulating factor (M-CSF), 50 ng/mL, receptor activator of NF kappaB ligand (RANKL) and 5 ng/mL transforming growth factor-β1 (TGF-β1) (Susa *et al.*, 2004; Agrawal *et al.*, 2011).

Alternatively peripheral blood mononuclear cells can be incubated with bone-derived stromal cells or osteoblast-like cells and under routine culture conditions osteoclast formation can be observed. MSCs have also been shown to induce osteoclast differentiation from mouse mononuclear phagocytes and monocytes in co-culture (Sabokbar and Athanasou, 2003). The addition of 1, 25-dihydroxyvitamin D_3 (1, 25-$(OH)_2D_3$) is known to aid the formation of osteoclast from these co-cultures (Sabokbar and Athanasou, 2003).

4.3.3 Stem cells

Cells of the osteoblast lineage can also be derived from sources such as MSCs and embryonic stem cells (ESCs). Understanding how these sources can be manipulated and induced to form mature cell phenotypes has become the subject of intense research activity and the cornerstone of many regenerative medicine treatment strategies. For musculoskeletal applications, MSCs isolated from sources such as bone marrow and adipose tissues can be expanded on plastic surfaces in 2D as undifferentiated cells. These *in vitro* expanded MSCs (Fig. 4.2) can then be differentiated *in vitro* towards the osteoblast lineage with the use of commercially available osteogenic differentiation medium such as OsteoPrime Induction Kit (PAA Laboratories GmbH, Austria) and OGM™ Osteoblast Growth Medium differentiation SingleQuots™ Kit (Lonza, Walkersville, MD, USA). Alternatively, in-house differentiation medium formulations that generally contain DMEM or

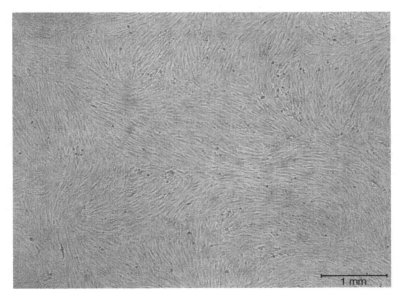

4.2 Phase contrast photomicrograph of human mesenchymal stem cells (hMSCs) showing their *in vitro* appearance (scale bar = 1 mm).

64 Standardisation in cell and tissue engineering

α-MEM-based medium supplemented with 10–20% (v/v) FBS, antibiotics and a 1000 mg/L glucose, 2 mM glutamine can be used and further supplemented with 10 nM dexamethasone, 0.1 mM L-ascorbic acid-2-phosphate, 5 mM β-glycerol phosphate (Pittenger *et al.*, 1999; Muller *et al.*, 2006). Differentiation of MSCs can also be induced with the addition of 100 ng/mL bone morphogenetic protein-2 (BMP-2) (Muller *et al.*, 2006).

ESCs such as H1 HESC (H1 human embryonic stem cells) can also be expanded in 2D on mitomycin-C inactivated mouse embryonic fibroblast feeder layers as undifferentiated ESCs. These ESCs are semi-differentiated in micromass culture and expanded in monolayer before being cultured in scaffolds with 10 μM ascorbate, 10 nM dexamethasone for 14 days (Bielby *et al.*, 2004). Osteogenic differentiation of ESCs have been reported to be improved with the use of the hanging drop method in combination with a time-optimised addition of 1α,25-(OH)$_2$ vitamin D$_3$, all-trans retinoic acid, and BMP-2, with increased end point efficiency of up to 90% (Kuske *et al.*, 2011).

4.4 Cell culture parameters: cartilage tissue culture

Within cartilage, chondrocytes are a heterogeneous population of cells exhibiting regional differences in the profile of matrix components that they synthesise and remodel. Cell shape and contact with ECM components are key drivers in the regulation of chondrocyte activity and as a consequence this should be considered when interpreting experimental data using plastic-adherent chondrocyte cell lines and primary cells.

4.4.1 Cell lines

A murine chondrocyte cell line (ATDC-5), human immortalised chondrocyte cell line (T/C-28a2) and human chondrosarcoma cell line (SW1353 and H-HEMC-SS) are used to model chondrocyte function *in vitro* (Bui *et al.*, 2010; Chao *et al.*, 2011; Conde *et al.*, 2011). ATDC-5 cells are routinely cultured in monolayer using the basal medium DMEM-Ham's F-12 medium with 5% (v/v) FBS, 10 ng/mL transforming growth factor beta 3 (TGFβ3) and 3×10^{-8} M sodium selenite or 1× insulin, transferrin and selenium (ITS) which contains 10 μg/mL insulin, 5.5 μg/mL transferrin, 6.7 ng/mL sodium selenite (Gomez *et al.*, 2009). T/C-28a2 and SW1353 cells have been cultured in the same basal medium supplemented with 10% (v/v) FBS, 2 mM L-glutamine and antibiotics (Gomez *et al.*, 2009; Chao *et al.*, 2011). These cell lines have been investigated in high-density monolayers and in 3D micromass cultures and maintained for approximately 4 weeks in culture to induce re-differentiation to a chondrocytic phenotype.

© Woodhead Publishing Limited, 2013

4.4.2 Primary cultures from isolated tissues

Primary chondrocytes can be isolated from several tissue sources. Hyaline/articular chondrocytes can be isolated from preserved cartilage areas of human patients undergoing knee and hip arthroplasty for osteoarthritis or trauma (Iannone *et al.*, 2002). Alternatively, bovine sternum, nasal, meta-carpophalangeal and other articular cartilage joints can be used as a source of chondrocytes. Also, porcine knee femoral condyle bone plugs and ovine tissue such as the facet joint and end plate cartilage are also useful cell sources.

Dedifferentiation of primary chondrocytes

Terminally differentiated human chondrocytes within cartilage exhibit limited ability to proliferate but in monolayer culture these chondrocytes can undergo dedifferentiation and regain significant proliferative capacity. Primary dedifferentiated chondrocytes are generally investigated in high-density monolayers and in 3D micromass cultures following isolation by enzymatic digestion (0.2% (w/v) collagenase for 3 h, 37°C and 0.02% (w/v) hyaluronidase for 30 min, 37°C and 0.25% (w/v) pronase for 90 min, 37°C) (von der Mark *et al.*, 1977; Iannone *et al.*, 2002) and *in vitro* expanded at 37°C in 5% CO_2 and 95% air.

Expansion of dedifferentiated chondrocytes

Dedifferentiated chondrocytes are usually cultured for periods of 28 days. This is the general reported time for the induction of re-differentiation to a chondrocytic phenotype. Dedifferentiated chondrocytes can be expanded in monolayer using commercially available culture media such as ChondroPrime S Kit and ChondroPrime SF Kit (serum and serum-free respectively) from company PAA GmbH (Austria) or CGM™ Chondrocyte Growth Media BulletKit™ and CDM™ Lonza (Walkersville, MD, USA). Chondrocyte differentiation can also be supported by media prepared in-house as follows. *Composition A*: DMEM with 10% (v/v) FCS, 1 µg/mL insulin and 50 µg/mL L-ascorbic acid. *Composition B*: used by Xu *et al.* (2006) consists of DMEM buffered with a combination of 44 mM $NaHCO_3$ and 25 mM HEPES at pH 7.4 and osmolarity adjusted to 380 mOsm. *Composition C*: DMEM supplemented with 10% (v/v) FBS, antibiotics and 2 mM L-glutamine (von der Mark *et al.*, 1977). *Composition D*: DMEM supplemented with 10% FBS, 1% of 1 × MEM, 1 × non-essential amino acids, without L-glutamine; 10 000 units/cm³ penicillin; 10 000 µg/cm³ streptomycin; 20 mM L-alanylglutamine and 10 mM 4-(2-hydroxyethyl)-1-piper-azineethanesulphonic acid (HEPES) buffer with 1 µL/cm³ basic fibroblast growth factor stock solution (bFGF or FGF-2; stock solution of 10 µg/cm³

66　Standardisation in cell and tissue engineering

human growth factor in phosphate-buffered saline (PBS) containing 1 mg/ cm^3 bovine serum albumin (BSA)) (Lapworth *et al.*, 2012).

Re-differentiation of dedifferentiated chondrocytes

For chondrocytes to produce ECM in an appropriate manner, they must be cultured in an *in vitro* 3D environment (e.g., cell micromass). Differentiation media such as Chondrocyte Differentiation Media BulletKit™ from Lonza (Walkersville, MD, USA) support this process. Alternatively in-house media recipes can be used that include 100 ng/mL FGF-2 (bFGF) and 10 ng/mL TGFβ1 in serum-free medium and ITS for up to 21 days (Khan *et al.*, 2011; Lapworth *et al.*, 2012); or the addition of 1 µL/cm^3 insulin stock solution from bovine pancreas, stock solution of 1 mg/cm^3 in 100 mM acetic acid and 1 µL/cm^3 ascorbic acid (50 mg/mL in DMEM) (Lapworth *et al.*, 2012). In addition to normoxic culture conditions of 20% oxygen partial pressure, hypoxic culture conditions of 5% have been studied and used in the propagation and re-differentiation of articular chondrocytes *in vitro* (Schrobback *et al.*, 2012).

4.4.3　Stem cells

ESCs and MSCs of human origin have been differentiated into chondrocytes with the long-term goal of their use for cartilage repair (Pittenger *et al.*, 1999; Mathur *et al.*, 2012). ESCs have been differentiated into chondrogenic lineages using BMP-2 and TGFβ1 with an approximate end point efficiency of 60% (Kuske *et al.*, 2011). To promote chondrogenesis MSCs are cultured to confluency before being harvested and seeded at high densities to form micromasses: (a) 0.25–0.5 × 10^6 cells in a 15 mL polypropylene tube are gently centrifuged to form a pelleted micromass or (b) approximately 200 000 cells/10 µL can be seeded and incubated for 2 h at the centre of a coverslip before topping up with either a commercial chondrogenic medium, for example, Chondrocyte Differentiation Media BulletKit™ from Lonza, Walkersville, MD, USA, supplemented with 10 ng/mL TGFβ3 (an inducing agent) or an in-house chondrogenic medium (DMEM supplemented with 2 mM L-glutamine, antibiotics, 100 nM dexamethasone, 50 µg/mL ascorbic acid-2-phosphate, 40 µg/mL proline, 1 × ITS + L premix and 10 ng/mL TGFβ3 (an inducing agent)) (Mackay *et al.*, 1998; Yoo *et al.*, 1998; Murdoch *et al.*, 2007). Micromasses are fed with fresh chondrogenic medium every 2–3 days and incubated at 37°C in a 5% CO_2 incubator. At 14–28 days in culture micromasses can be harvested and studied further (Mackay *et al.*, 1998; Giovannini *et al.*, 2010). It's been reported that these cultures can be further differentiated to the hypertrophic state by the addition of 50 nM

© Woodhead Publishing Limited, 2013

thyroxine, the withdrawal of TGFβ3 and the reduction of dexamethasone to 1 nM (Mackay *et al.*, 1998).

4.5 Two-dimensional tissue culture methods for hard tissues

Much of our understanding of the behaviour of bone and cartilage cells has been achieved using 2D culture surfaces, predominantly tissue culture grade plastic ware or glass. The exception to this is the study of the activity of osteoclasts which has been investigated by assessing these cells on mineralised matrix, for example, devitalised cortical bone or dentine.

In order to study the role that the extracellular environment may play in regulating the activity of musculoskeletal cells, proteins and other biomolecules have been either passively absorbed or covalently immobilised prior to cell culture experimentation (Rezania and Healy, 1999; Huang *et al.*, 2003; Kirkwood *et al.*, 2003). In refinement of these approaches, proteins can be engineered to contain a terminal cysteine residue as part of scaffold protein to form an ordered self-assembled monolayer (SAM) on a gold surface. Peptide and protein motifs can be engineered into the scaffold protein and therefore this methodology provides a durable immobilisation strategy to investigate their effect on cell activity (Mitchell *et al.*, 2010). SAM approaches have also been used to display discrete chemical functionalities and assess their role in the regulation of cell activity (Nakaoka *et al.*, 2010).

Regional control of the distribution of proteins and other ligands on a surface can be achieved using a soft lithography approach (Fig. 4.3). In this technique a mould is fabricated, most often using a lithographic process to create a patterned silicon wafer. A patterned stamp is then created by casting a polymer such as polydimethylsiloxane (PDMS) in this mould. Once cured the PDMS can be 'inked' with the ligands of interest and then transferred to a cell-culture surface by simple contact transfer (Xia and Whitesides, 1998). This approach allows peptides, proteins, other biomolecules and chemicals to be organised on surfaces in discrete regional patterns facilitating studies on their influence on cell adhesion, morphology and control of cell phenotype.

A more recently exploited approach to the controlled fabrication of surfaces is the use of a layer-by-layer methodology (Lavalle *et al.*, 2011). In this way it is possible to coat surfaces by successive rounds of exposure to polyelectrolytes with complimentary charges (Zhang *et al.*, 2011). This approach not only creates surfaces that support cell adhesion and exhibit time-dependent release of growth factors or therapeutic agents (Volodkin *et al.*, 2011) but can also be used to subtly influence the mechanical properties of surfaces to allow study of the effect that this may have on cell activity (Kocgozlu *et al.*, 2010).

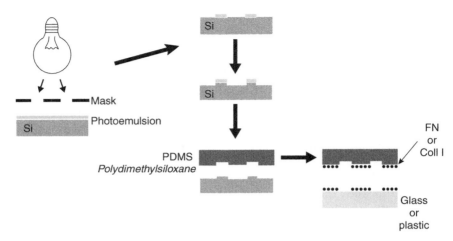

4.3 Schematic illustrating the process flow for soft lithography. A mould is created using lithographic techniques in a silicon wafer. A PDMS reproduction of the mould is created and then used in a tissue culture environment to create immobilised patterns of biomolecules such as collagen of fibronectin.

4.6 Two-and-a-half- and three-dimensional tissue culture methods for hard tissues

Many tissue engineering approaches target the musculoskeletal system and as a consequence there has been a growing emphasis on understanding how bone and cartilage cells behave in 3D scaffolds. Analysis of these scaffolds is often pursued at first *in vitro* in an attempt to describe how 3D structure at the nano, micro and macro scale can be manipulated to influence cell phenotype (Mathur *et al.*, 2012).

4.6.1 Bone

It has been well established that 3D culture environments profoundly influence the commitment of MSCs, their subsequent differentiation to the osteoblast lineage as well as the activity of mature osteoblasts. It is beyond the remit of this chapter to describe the fabrication strategies for all of these scaffolds in detail but some examples of the materials that help to provide insight into the underlying biology of musculoskeletal cells are outlined.

Experimental approaches that give surfaces with defined topography are considered by some to have 2½D. These provide a useful tool to begin to explore the way in which surface mediated effects can influence cell function. A variety of approaches can be used to fabricate these surfaces,

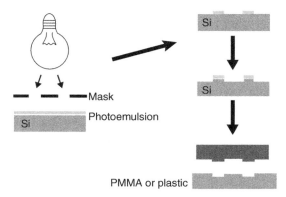

4.4 Schematic illustrating the process flow for hot embossing. A template is created by etching a silicon wafer following lithographic processes. The silicon template is then pressed into a substrate such as polycarbonate or polymethylmethacrylate (PMMA) that has been heated to above its glass transition temperature. The two are cooled and eased apart to allow the use of the patterned plastic as a culture substrate.

including grit blasting and physical abrasion through to the creation of highly controlled patterned surfaces with the use of reproduction methodologies such as hot embossing (Fig. 4.4) or ion beam lithography (Falconnet *et al.*, 2006).

The simplest 3D environment in which to culture cells is to encase them in biopolymers such as type I collagen or fibrin (Catelas *et al.*, 2006). Another approach is to use peptides (e.g., RAD16-I) that self-assemble into fibrillar structures that are ECM-like and support cell attachment, growth and differentiation (Hauser and Zhang, 2010). An alternative is to combine chemical polymers, for example, polyethylene-based with peptides or protein motifs to create a cross-linked hydrogel that is capable of providing a tailored molecular network to control cell growth (Kloxin *et al.*, 2010).

The environment created by biological and chemical polymers, either alone or in combination with peptide or protein motifs (as stated above), is often mechanically weak in comparison to that found in naturally occurring bone. For this reason attempts have been made to investigate how macroporous scaffolds created by various approaches, including the casting of polymers around salt granules (Chiu *et al.*, 2010) or under high internal phase emulsion conditions (Akay *et al.*, 2004), bioceramics (Takagi *et al.*, 2010) or porous metals (Yang *et al.*, 2011), support osteoblast activity. This approach can be combined with the hydrogels described above to create composite scaffolds that display appropriate bulk mechanical properties and highly controlled cellular environments (Bokhari *et al.*, 2005).

70   Standardisation in cell and tissue engineering

Culturing of cells on a 3D scaffold is challenging because sufficient exchange of gases, nutrients and waste products have to be maintained in order to ensure that cell viability is not compromised within the material. Under static culture conditions cell viability is limited at depths of 0.5–1.0 mm (depending on the organisation of the scaffold) (Cartmell *et al.*, 2003). For this reason a range of flow and bioreactor culture systems have been developed to maintain oxygenation and metabolite exchange throughout a cell-seeded scaffold. This area of cell culture is highly specialised; therefore more exhaustive commentary can be found elsewhere (Rauh *et al.*, 2011).

In addition, to improve mechanical properties of engineered hard tissue, studies have been performed with the application of mechanical loads such as stress and strain at different magnitudes. These mechanical loads are known to affect bone formation and resorption (Kumbar *et al.*, 2011).

4.6.2 Cartilage

The seeding of cells at high density in 3D is associated with enhanced tissue formation. 3D organisation is central to the regulation of the chondrocyte phenotype and the simplest of 3D tissue culture methods used is aggregating cultures, for example, chondrogenic micromass formation (pelleting of 5×10^5 cells in 15 mL tube by centrifugation at $500 \times g$ for 5 min at 10°C) or the use of rotational culture systems in the generation of articular-like cartilage (Furukawa *et al.*, 2003; Giovannini *et al.*, 2008; Giovannini *et al.*, 2010).

Chondrocytes isolated from bovine articular cartilage have been cultured on 1.2% alginate beads and agarose scaffold for 12 days at a seeding density of 4 000 000 cells/mL. Under perfusion conditions chondrocytes maintained a differentiated morphology within alginate beads resulting in the production and retention of more glycosaminoglycans (GAGs) (Xu *et al.*, 2006). Bovine chondrocytes have also been embedded into agarose gels and cultured at a controlled extracellular pH 7.2 which maximised ECM synthesis, for example, GAGs production which is an important parameter for successful cartilage tissue engineering (Wu *et al.*, 2007). The application of cyclic compressions (0–15%, at a frequency of 1.0 Hz) for 0.5–48 h has also been used on 3D constructs such as chondrocyte/agarose (McGlashan *et al.*, 2010). To further increase tissue formation and to improve mechanical properties of tissue-engineered cartilage, mechanical stimulation of constructs, that is, application of tensile strain and fluid flow have been studied (Mawatari *et al.*, 2010). In addition, the application of intermittent static biaxial tensile strains has been investigated for their effects on the activity of chondrocytes *in vitro* and the resulting deposition of matrix (Fan and Waldman, 2010). In addition, chondrogenesis has been induced with the use of MSCs treated with TGFβ3 in alginate bead systems (Lee *et al.*, 2004).

© Woodhead Publishing Limited, 2013

4.7 Conclusion

Tissue culture approaches have led to a deeper understanding of the molecular mechanisms that control the activity of bone and cartilage cells. This has encompassed not only the explanation of the role that a number of hormones, cytokines and growth factors play in the normal and pathobiology of these tissues but also a growing recognition of the significance of the extracellular environment and its organisation. The promise of advances in cell therapy and tissue engineering means that there will be a continued impetus to develop and refine the *in vitro* approaches described here to support the growth of bone and cartilage cells.

Culture methods described in this chapter are approaches adopted for laboratory research investigating and understanding the activity of bone and cartilage cells. For translation of these cell types as treatments in clinical practice there are defined regulatory standards that need to be followed in terms of their culture and transplantation. The growth of cells for use as part of a treatment requires distinct and standardised protocols with record-keeping at the heart of the process. In addition, approved reagents need to be used including plasticware and the avoidance of animal-derived supplements or factors. Regulatory bodies governing these protocols in the United Kingdom (UK) include but are not limited to the Human Tissue Authority (HTA), which governs protocols and usage of advanced therapy medical products (ATMPs). The HTA definition of an ATMP identifies that it:

> Contains or consists of cells or tissues that have either been subject to 'substantial manipulation' or that are not intended to be used for the same essential function(s) in the recipient as in the donor; and is presented as having properties for treating or preventing disease in patients (http://www.hta.gov.uk/licensingandinspections/sectorspecificinformation/tissueandcellsforpatienttreatment/advancedtherapymedicinalproductsfaqs.cfm).

In addition, the Department of Health within the UK governs protocols and the use of stem cells (e.g., hMSCs and hESCs) and their derivatives via the regulatory tool, UK Stem Cell Tool Kit.

4.8 References

Agrawal, A., Gallagher, J. A. and Gartland, A. (2011) Human osteoclast culture and phenotypic characterization. *Methods Mol Biol*, **806**, 357–375.

Akay, G., Birch, M. A. and Bokhari, M. A. (2004) Microcellular polyHIPE polymer supports osteoblast growth and bone formation *in vitro*. *Biomaterials*, **25**, 3991–4000.

Bakker, A. and Klein-Nulend, J. (2003) Osteoblast isolation from murine calvariae and long bones. *Methods Mol Med*, **80**, 19–28.

72 Standardisation in cell and tissue engineering

Bielby, R. C., Boccaccini, A. R., Polak, J. M. and Buttery, L. D. (2004) In vitro differentiation and *in vivo* mineralization of osteogenic cells derived from human embryonic stem cells. *Tissue Eng*, **10**, 1518–1525.

Bliziotes, M., Eshleman, A., Burt-Pichat, B., Zhang, X. W., Hashimoto, J., Wiren, K. and Chenu, C. (2006) Serotonin transporter and receptor expression in osteocytic MLO-Y4 cells. *Bone*, **39**, 1313–1321.

Bokhari, M. A., Akay, G., Zhang, S. G. and Birch, M. A. (2005) Enhancement of osteoblast growth and differentiation in vitro on a peptide hydrogel – poly-HIPE polymer hybrid material. *Biomaterials*, **26**, 5198–5208.

Bonewald, L. F. (1999) Establishment and characterization of an osteocyte-like cell line, MLO-Y4. *J Bone Miner Metab*, **17**, 61–65.

Bui, C., Ouzzine, M., Talhaoui, I., Sharp, S., Prydz, K., Coughtrie, M. W. and Fournel-Gigleux, S. (2010) Epigenetics: methylation-associated repression of heparan sulfate 3-O-sulfotransferase gene expression contributes to the invasive phenotype of H-EMC-SS chondrosarcoma cells. *FASEB J*, **24**, 436–450.

Cartmell, S. H., Porter, B. D., Garcia, A. J. and Guldberg, R. E. (2003) Effects of medium perfusion rate on cell-seeded three-dimensional bone constructs *in vitro*. *Tissue Eng*, **9**, 1197–1203.

Catelas, I., Sese, N., Wu, B. M., Dunn, J. C. Y., Helgerson, S. and Tawil, B. (2006) Human mesenchymal stem cell proliferation and osteogenic differentiation in fibrin gels *in vitro*. *Tissue Eng*, **12**, 2385–2396.

Chan, M. E., Lu, X. L., Huo, B., Baik, A. D., Chiang, V., Guldberg, R. E., Lu, H. H. and Guo, X. E. (2009) A trabecular bone explant model of osteocyte-osteoblast co-culture for bone mechanobiology. *Cell Mol Bioeng*, **2**, 405–415.

Chao, P. Z., Hsieh, M. S., Cheng, C. W., Lin, Y. F. and Chen, C. H. (2011) Regulation of MMP-3 expression and secretion by the chemokine eotaxin-1 in human chondrocytes. *J Biomed Sci*, **18**, 86.

Chiu, Y.-C., Larson, J. C., Isom, A. Jr. and Brey, E. M. (2010) Generation of porous poly(ethylene glycol) hydrogels by salt leaching. *Tissue Eng Pt C Meth*, **16**, 905–912.

Collin-Osdoby, P. and Osdoby, P. (2012) RANKL-mediated osteoclast formation from murine RAW 264.7 cells. *Methods Mol Biol*, **816**, 187–202.

Conde, J., Gomez, R., Bianco, G., Scotece, M., Lear, P., Dieguez, C., Gomez-Reino, J., Lago, F. and Gualillo, O. (2011) Expanding the adipokine network in cartilage: identification and regulation of novel factors in human and murine chondrocytes. *Ann Rheum Dis*, **70**, 551–559.

Costa-Rodrigues, J., Fernandes, A. and Fernandes, M. H. (2011) Reciprocal osteoblastic and osteoclastic modulation in co-cultured MG63 osteosarcoma cells and human osteoclast precursors. *J Cell Biochem*, **112**, 3704–3713.

Falconnet, D., Csucs, G., Grandin, H. M. and Textor, M. (2006) Surface engineering approaches to micropattern surfaces for cell-based assays. *Biomaterials*, **27**, 3044–3063.

Fan, J. C. and Waldman, S. D. (2010) The effect of intermittent static biaxial tensile strains on tissue engineered cartilage. *Ann Biomed Eng*, **38**, 1672–1682.

Furukawa, K. S., Suenaga, H., Toita, K., Numata, A., Tanaka, J., Ushida, T., Sakai, Y. and Tateishi, T. (2003) Rapid and large-scale formation of chondrocyte aggregates by rotational culture. *Cell Transplant*, **12**, 475–479.

© Woodhead Publishing Limited, 2013

2- and 3-D tissue culture methods for hard tissue engineering   73

Gallagher, J. A., Gundle, R. and Beresford, J. N. (1996) Isolation and culture of bone-forming cells (osteoblasts) from human bone. *Methods Mol Med*, **2**, 233–262.

Gartland, A., Rumney, R. M., Dillon, J. P. and Gallagher, J. A. (2012) Isolation and culture of human osteoblasts. *Methods Mol Biol*, **806**, 337–355.

Giovannini, S., Brehm, W., Mainil-Varlet, P. and Nesic, D. (2008) Multilineage differentiation potential of equine blood-derived fibroblast-like cells. *Differentiation*, **76**, 118–129.

Giovannini, S., Diaz-Romero, J., Aigner, T., Heini, P., Mainil-Varlet, P. and Nesic, D. (2010) Micromass co-culture of human articular chondrocytes and human bone marrow mesenchymal stem cells to investigate stable neocartilage tissue formation *in vitro*. *Eur Cell Mater*, **20**, 245–259.

Gomez, R., Lago, F., Gomez-Reino, J. J., Dieguez, C. and Gualillo, O. (2009) Expression and modulation of ghrelin O-acyltransferase in cultured chondrocytes. *Arthritis Rheum*, **60**, 1704–1709.

Han, S. H., Kim, K. H., Han, J. S., Koo, K. T., Kim, T. I., Seol, Y. J., Lee, Y. M., Ku, Y. and Rhyu, I. C. (2011) Response of osteoblast-like cells cultured on zirconia to bone morphogenetic protein-2. *J Periodontal Implant Sci*, **41**, 227–233.

Hauser, C. A. E. and Zhang, S. (2010) Designer self-assembling peptide nanofiber biological materials. *Chem Soc Rev*, **39**, 2780–2790.

Hefley, T., Cushing, J. and Brand, J. S. (1981) Enzymatic isolation of cells from bone: cytotoxic enzymes of bacterial collagenase. *Am J Physiol*, **240**, C234–C238.

Hillsley, M. V. (1999) Methods to isolate, culture, and study osteoblasts. *Methods Mol Med*, **18**, 293–301.

Huang, H., Zhao, Y., Liu, Z., Zhang, Y., Zhang, H., Fu, T. and Ma, X. (2003) Enhanced osteoblast functions on RGD immobilized surface. *J Oral Implant*, **29**, 73–79.

Iannone, F., De Bari, C., Dell'accio, F., Covelli, M., Patella, V., Lo Bianco, G. and Lapadula, G. (2002) Increased expression of nerve growth factor (NGF) and high affinity NGF receptor (p140 TrkA) in human osteoarthritic chondrocytes. *Rheumatology (Oxford)*, **41**, 1413–1418.

Javed, A., Chen, H. and Ghori, F. Y. (2010) Genetic and transcriptional control of bone formation. *Oral Maxillofac Surg Clin North Am*, **22**, 283–293, v.

Kato, Y., Windle, J. J., Koop, B. A., Mundy, G. R. and Bonewald, L. F. (1997) Establishment of an osteocyte-like cell line, MLO-Y4. *J Bone Miner Res*, **12**, 2014–2023.

Khan, I. M., Evans, S. L., Young, R. D., Blain, E. J., Quantock, A. J., Avery, N. and Archer, C. W. (2011) Fibroblast growth factor 2 and transforming growth factor beta1 induce precocious maturation of articular cartilage. *Arthritis Rheum*, **63**, 3417–3427.

Kirkwood, K., Rheude, B., Kim, Y. J., White, K. and Dee, K. C. (2003) *In vitro* mineralization studies with substrate-immobilized bone morphogenetic protein peptides. *J Oral Implant*, **29**, 57–65.

Kloxin, A. M., Tibbitt, M. W. and Anseth, K. S. (2010) Synthesis of photodegradable hydrogels as dynamically tunable cell culture platforms. *Nat Prot*, **5**, 1867–1887.

Kocgozlu, L., Lavalle, P., Koenig, G., Senger, B., Haikel, Y., Schaaf, P., Voegel, J.-C., Tenenbaum, H. and Vautier, D. (2010) Selective and uncoupled role of substrate elasticity in the regulation of replication and transcription in epithelial cells. *J Cell Sci*, **123**, 29–39.

Kumbar, S. G., Toti, U. S., Deng, M., James, R., Laurencin, C. T., Aravamudhan, A., Harmon, M. and Ramos, D. M. (2011) Novel mechanically competent polysaccharide scaffolds for bone tissue engineering. *Biomed Mater*, **6**, 065005.

Kuske, B., Savkovic, V. and Zur Nieden, N. I. (2011) Improved media compositions for the differentiation of embryonic stem cells into osteoblasts and chondrocytes. *Methods Mol Biol*, **690**, 195–215.

Lapworth, J. W., Hatton, P. V., Goodchild, R. L. and Rimmer, S. (2012) Thermally reversible colloidal gels for three-dimensional chondrocyte culture. *J R Soc Interface*, **9**(67), 362–375.

Lavalle, P., Voegel, J.-C., Vautier, D., Senger, B., Schaaf, P. and Ball, V. (2011) Dynamic aspects of films prepared by a sequential deposition of species: perspectives for smart and responsive materials. *Adv Mater*, **23**, 1191–1221.

Lee, J. W., Kim, Y. H., Kim, S. H., Han, S. H. and Hahn, S. B. (2004) Chondrogenic differentiation of mesenchymal stem cells and its clinical applications. *Yonsei Med J*, **45**(Suppl), 41–47.

Mackay, A. M., Beck, S. C., Murphy, J. M., Barry, F. P., Chichester, C. O. and Pittenger, M. F. (1998) Chondrogenic differentiation of cultured human mesenchymal stem cells from marrow. *Tissue Eng*, **4**, 415–428.

Mathur, D., Pereira, W. C. and Anand, A. (2012) Emergence of chondrogenic progenitor stem cells in transplantation biology- prospects and drawbacks. *J Cell Biochem*, **113**(2), 397–403.

Mawatari, T., Lindsey, D. P., Harris, A. H., Goodman, S. B., Maloney, W. J. and Smith, R. L. (2010) Effects of tensile strain and fluid flow on osteoarthritic human chondrocyte metabolism *in vitro*. *J Orthop Res*, **28**, 907–913.

McGlashan, S. R., Knight, M. M., Chowdhury, T. T., Joshi, P., Jensen, C. G., Kennedy, S. and Poole, C. A. (2010) Mechanical loading modulates chondrocyte primary cilia incidence and length. *Cell Biol Int*, **34**, 441–446.

Mitchell, E. A., Chaffey, B. T., Mccaskie, A. W., Lakey, J. H. and Birch, M. A. (2010) Controlled spatial and conformational display of immobilised bone morphogenetic protein-2 and osteopontin signalling motifs regulates osteoblast adhesion and differentiation *in vitro*. *BMC Biology*, **8**, 57.

Muller, I., Kordowich, S., Holzwarth, C., Spano, C., Isensee, G., Staiber, A., Viebahn, S., Gieseke, F., Langer, H., Gawaz, M. P., Horwitz, E. M., Conte, P., Handgretinger, R. and Dominici, M. (2006) Animal serum-free culture conditions for isolation and expansion of multipotent mesenchymal stromal cells from human BM. *Cytotherapy*, **8**, 437–444.

Murdoch, A. D., Grady, L. M., Ablett, M. P., Katopodi, T., Meadows, R. S. and Hardingham, T. E. (2007) Chondrogenic differentiation of human bone marrow stem cells in transwell cultures: generation of scaffold-free cartilage. *Stem Cells*, **25**, 2786–2796.

Nakaoka, R., Yamakoshi, Y., Isama, K. and Tsuchiya, T. (2010) Effects of surface chemistry prepared by self-assembled monolayers on osteoblast behavior. *J Biomed Mater Res Part A*, **94A**, 524–532.

Nijweide, P. J., van der Plas, A., Alblas, M. J. and Klein-Nulend, J. (2003) Osteocyte isolation and culture. *Methods Mol Med*, **80**, 41–50.

Owen, T. A. and Pan, L. C. (2008) Isolation and culture of rodent osteoprogenitor cells. *Methods Mol Biol*, **455**, 3–18.

Pittenger, M. F., Mackay, A. M., Beck, S. C., Jaiswal, R. K., Douglas, R., Mosca, J. D., Moorman, M. A., Simonetti, D. W., Craig, S. and Marshak, D. R. (1999) Multilineage potential of adult human mesenchymal stem cells. *Science*, **284**, 143–147.

Rauh, J., Milan, F., Guenther, K.-P. and Stiehler, M. (2011) Bioreactor systems for bone tissue engineering. *Tissue Eng Pt B Rev*, **17**, 263–280.

Rezania, A. and Healy, K. E. (1999) Biomimetic peptide surfaces that regulate adhesion, spreading, cytoskeletal organization, and mineralization of the matrix deposited by osteoblast-like cells. *Biotechnol Prog*, **15**, 19–32.

Sabokbar, A. and Athanasou, N. S. (2003) Generating human osteoclasts from peripheral blood. *Methods Mol Med*, **80**, 101–111.

San Miguel, S. M., Fatahi, M. R., Li, H., Igwe, J. C., Aguila, H. L. and Kalajzic, I. (2009) Defining a visual marker of osteoprogenitor cells within the periodontium. *J Periodontal Res*, **45**, 60–70.

Sauren, Y. M., Mieremet, R. H., Groot, C. G. and Scherft, J. P. (1989) An electron microscopical study on the presence of proteoglycans in the calcified bone matrix by use of cuprolinic blue. *Bone*, **10**, 287–294.

Schrobback, K., Klein, T. J., Crawford, R., Upton, Z., Malda, J. and Leavesley, D. I. (2012) Effects of oxygen and culture system on *in vitro* propagation and redifferentiation of osteoarthritic human articular chondrocytes. *Cell Tissue Res*, **347**(3), 649–663.

Semeins, C. M., Bakker, A. D. and Klein-Nulend, J. (2012) Isolation of primary avian osteocytes. *Methods Mol Biol*, **816**, 43–53.

Susa, M., Luong-Nguyen, N. H., Cappellen, D., Zamurovic, N. and Gamse, R. (2004) Human primary osteoclasts: *in vitro* generation and applications as pharmacological and clinical assay. *J Transl Med*, **2**, 6.

Takagi, K., Takahashi, T., Kikuchi, K. and Kawasaki, A. (2010) Fabrication of bioceramic scaffolds with ordered pore structure by inverse replication of assembled particles. *J Euro Ceram Soc*, **30**, 2049–2055.

Volodkin, D., Skirtach, A. and Moehwald, H. (2011) LbL films as reservoirs for bioactive molecules. In: Borner, H.G., Börner, H. G. and Lutz, J-F. (eds.) *Bioactive Surfaces*, Springer.

von der Mark, K., Gauss, V., von der Mark, H. and Muller, P. (1977) Relationship between cell shape and type of collagen synthesised as chondrocytes lose their cartilage phenotype in culture. *Nature*, **267**, 531–532.

Wei, S., Teitelbaum, S. L., Wang, M. W. and Ross, F. P. (2001) Receptor activator of nuclear factor-kappa b ligand activates nuclear factor-kappa b in osteoclast precursors. *Endocrinology*, **142**, 1290–1295.

Westbroek, I., De Rooij, K. E. and Nijweide, P. J. (2002) Osteocyte-specific monoclonal antibody MAb OB7.3 is directed against Phex protein. *J Bone Miner Res*, **17**, 845–853.

Wu, M. H., Urban, J. P., Cui, Z. F., Cui, Z. and Xu, X. (2007) Effect of extracellular ph on matrix synthesis by chondrocytes in 3D agarose gel. *Biotechnol Prog*, **23**, 430–434.

Xia, Y. N. and Whitesides, G. M. (1998) Soft lithography. *Angewandte Chemie-International Edition*, **37**, 551–575.

Xu, X., Urban, J. P., Tirlapur, U., Wu, M. H. and Cui, Z. (2006) Influence of perfusion on metabolism and matrix production by bovine articular chondrocytes in hydrogel scaffolds. *Biotechnol Bioeng*, **93**, 1103–1111.

Yamamoto, N., Furuya, K. and Hanada, K. (2002) Progressive development of the osteoblast phenotype during differentiation of osteoprogenitor cells derived from fetal rat calvaria: model for in vitro bone formation. *Biol Pharm Bull*, **25**, 509–515.

Yang, D., Shao, H., Guo, Z., Lin, T. and Fan, L. (2011) Preparation and properties of biomedical porous titanium alloys by gelcasting. *Biomed Mater*, **6**, 045010.

Yoo, J. U., Barthel, T. S., Nishimura, K., Solchaga, L., Caplan, A. I., Goldberg, V. M. and Johnstone, B. (1998) The chondrogenic potential of human bone-marrow-derived mesenchymal progenitor cells. *J Bone Joint Surg Am*, **80**, 1745–1757.

Yu, V. W., Akhouayri, O. and St-Arnaud, R. (2009) FIAT is co-expressed with its dimerization target ATF4 in early osteoblasts, but not in osteocytes. *Gene Expr Patterns*, **9**, 335–340.

Zhang, J.-T., Nie, J., Muehlstaedt, M., Gallagher, H., Pullig, O. and Jandt, K. D. (2011) Stable extracellular matrix protein patterns guide the orientation of osteoblast-like cells. *Adv Funct Mater*, **21**, 4079–4087.

5
Vascularisation of tissue-engineered constructs

B. BURANAWAT, P. KALIA and L. DI SILVIO,
King's College London, Dental Institute, UK

DOI: 10.1533/9780857098726.1.77

Abstract: A variety of biomaterials have been fabricated to replace tissues, either as an off-the-shelf option or as a cell-seeded construct. The formation of a mature vascular supply throughout such materials is vital to the success of its integration into host tissue, its long-term survival and regeneration of a mature, functional tissue. In this chapter, the development of healthy vasculature is reviewed in the context of tissue engineering. The importance of a healthy blood supply for bone growth and repair is discussed, as well as various pathologies associated with uncontrolled or dysfunctional blood supply, including tumorigenesis. A number of growth factors essential to healthy tissue formation and angiogenesis have been tested and may be used alone, or in combination with biomaterials and either a single cell type or multicellular co-cultures. Stem cell, osteoblast and endothelial cell co-culture techniques were optimised for tissue engineering of vascularised scaffolds, and vessel formation observed.

Key words: bone regeneration, blood supply, vasculogenesis, angiogenesis, tissue engineering, biomaterials, growth factors.

5.1 Introduction

Over the last few decades scientists and clinicians have worked together to adapt, develop and clinically use biomaterials either 'off the shelf' or seeded with appropriate cells and growth factors (GFs). Different strategies have been developed, each of which offers vast potential. However, the success of any *ex vivo* graft is highly dependent on vascular invasion within the graft to ensure adequate oxygen and nutrient supply. New bone formation within the graft will only occur if there is a rapid integration with the host vasculature. Without a functional vascular supply, a tissue-engineered construct will ultimately be prone to failure. The development of a mature vasculature is dependent on numerous factors: the migration of endothelial cells and other cells, an extracellular matrix (ECM) to provide structural support and growth factors to act as stimulating factors. For individual cells to survive *in situ*, their basic

78   Standardisation in cell and tissue engineering

requirements need to be met; these include a simultaneous supply of oxygen and nutrients along with an adequate removal of waste products. Delivery of growth factors and cytokines to surrounding cells occurs by virtue of diffusion to and from the network of capillaries which run throughout the tissue. Diffusion between cells and capillaries is a limiting factor, controlling how far a tissue can grow before new blood vessel formation is required. The supply of oxygen and nutrients to the tissue-engineering construct is often limited by diffusion processes that can only supply cells in a proximity of 100–200 µm from the next capillary.[1] *In vitro,* a solution exists that allows one to partially overcome this 'diffusion limit' through the use of a perfusion bioreactor.[2] However, in order for the implant to survive *in vivo,* the tissue has to be vascularised, meaning that a new capillary network capable of delivering nutrients to the cell is formed within the construct. The spontaneous vascular ingrowth is usually limited to several tenths of micrometres per day;[3] this is often seen as a part of an inflammatory wound healing response. Simultaneously, the growth of cells into and a rapid neovascularisation of the implant material must take place. Nevertheless, in larger tissue-engineered constructs, in order to surpass this tissue thickness limit and repair large, dynamic structures such as bone, achieving pre-existing vasculature or accelerated vascularisation is critical. Within the field of microvascular surgery there has been much long-term success in re-vascularising tissues. This provided a method for transplanting living bone segments with maintained blood supply. Incorporation and survival of vascularised grafts can therefore be expected to be superior over free grafts. However, there are also disadvantages such as morbidity and potential deformity of the donor site; furthermore, the technique necessitates a longer operation time. A critical developmental breakthrough will rely upon the ability to produce a vessel that adapts to physiological pressures and roles. No synthetic material can meet these requirements, and the success of any graft is dependent on a number of processes, the most important of which is vascularisation. A major problem for the reconstruction of large bone defects is the provision of a prompt vasculature, allowing for recruitment and survival of appropriate cell types. Many strategies for enhancing vascularisation are currently under investigation. In this chapter we discuss the potential for '*in vitro* prevascularisation' of a bone graft with the aim of creating instantaneous perfusion of a construct. Scaffold design, culture conditions and cell sources all play a significant role. Ideally, a graft that has been prevascularised *ex vivo* would be transplanted *in vivo* and encouraged to anastomose to the host vasculature. This type of approach could dramatically decrease the time that is needed to vascularise an entire construct, because host vessels only have to grow into its outer regions, that is, until the ingrowing vessels meet the preformed vascular network.[1] Prevascularisation generally refers to the formation of a well-connected capillary or microvessel network within an implantable tissue prior to implantation.[4]

© Woodhead Publishing Limited, 2013

5.2 Growth of healthy vessels – embryonic vasculogenesis

Development of the body's vasculature, from embryo to adult, begins in the yolk sac and trophoblast where 'blood islands', or haemangioblasts, first arise from aggregates of mesenchyme in the splanchnopleuric mesoderm, during the time of somite formation. Blood islands consist of haematopoetic precursor cells in the interior, and endothelial cells at the periphery.[3,5–10] A subset of the primitive mesodermal cells is commited towards differentiation into endothelial cells (ECs). It is thought that at this stage, molecules from the fibroblast growth factor (FGF) family, such as transforming growth factor (TGF-β), induce mesoderm formation.[11–13] The endothelial precursors are then thought to differentiate *in situ*.[14] Endothelial cells are believed to originate from haemangioblasts, a process termed 'primary angiogenesis'.[15] The endothelial cells in the blood islands then elongate and migrate, contacting adjacent blood islands and eventually fusing to produce the primary vascular plexus.[16] Both angiogenesis and vasculogenesis are involved in the embryonic development that follows. Later in embryo genesis as well as in adults, blood vessels originate from pre-existing endothelial cells in capillaries and pre- and post-capillary vessels,[17–19] a process termed 'secondary angiogenesis'.[15] Permeability of vessels has been shown to decrease significantly in chorioallantoic membrane vessels as soon as day 5 of embryo development, although angiogenesis is thought to proceed until day 12.[20–22] Vessel permeability is thought to be affected by vascular endothelial growth factor (VEGF) which is discussed in the next section; unusually high vascular permeability has been described in tumour growth (see 5.3.3, below).

5.2.1 Formation of microvascular networks

Unlike the thick vascular walls seen in arteries and veins, a capillary is composed of endothelial cells (ECs), basal membrane and pericytes. The ECs lining the innermost layer of any vasculature are also major regulators of cardiovascular physiology, the control of blood flow and vessel tone, as well as the thrombo-resistance and modulation of leukocyte interactions. Physiologically the formation of blood vessels can be classified in two categories: 1) vasculogenesis, the *de novo* formation from undifferentiated endothelial cells; or 2) by 'endothelial progenitor cells' (EPCs), or angiogenesis, the sprouting of new blood vessels from existing blood vessels. Until recently, angiogenesis was considered to be the only method by which new vessel formation could occur within adults, with vasculogenesis being restricted to embryogenesis. However, identification by Ashara *et al.* of EPCs circulating within peripheral blood has led to the acceptance that vasculogenesis is also a process that occurs within adults in conjunction with angiogenesis[23] (Fig. 5.1).

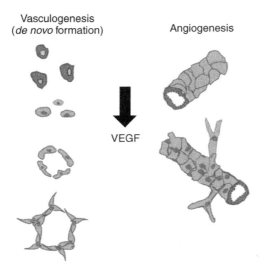

5.1 Formation of microvasculature. Primitive endothelial tubes are generated through vasculogenesis and angiogenesis. Vasculogenesis is initiated by the differentiation of progenitors to form angioblast-endothelial cells which then coalesce to form aggregates and lumen. VEGF promotes EC proliferation and also facilitates sprout formation.

Vasculogenesis

Vasculogenesis encompasses two main aspects: first, the primary differentiation of mesenchymal precursors (angioblasts) into endothelial cells; and second, the organisation of these cells to form capillary-like tubules. Ultimately the process of vasculogenesis is similar to that of angiogenesis. The major difference is that these *de novo* vessels are distinct from the pre-existing vascular supply, therefore anastomosis must occur in order to attain blood flow. Vasculogenesis is a multiphase process. Initially the endothelial cells must form the vessel primordial through aggregation and establishment of cell–cell contact. From here nascent endothelial tubes are formed which anastomose to produce a primary vascular network. Like angiogenesis, at this point the vessel walls are solely composed of endothelial cells, therefore recruitment and incorporation of mural cells (pericytes and smooth muscle) are required to stabilise the vessel.[24]

Angiogenesis

When quiescent, endothelial cells have a slow turnover rate due to cell-to-cell contact and inhibition of proliferation. However, in an activated state, for instance during wound healing, inflammation, ischaemia and within female reproductive organs, endothelial cells become active and change their phenotype to initiate angiogenesis. These changes can be divided into two main phases: the activation phase and the resolution phase. The activation phase consists of 4 basic stages, under tight control by a complex signalling cascade. (1) Vasodilatation of the parental vessel increases vascular permeability through a reduction in the contact between adjacent ECs, leading to increased extravascular fibrin deposition. (2) The basement membrane of a parental vessel is degraded through secretion and activation of a wide range of proteolytic enzymes. (3) This allows the migration of endothelial cells and proliferation to form a leading edge of the new capillary. (4) Finally, the capillary lumen and tube-like structure are formed. During the resolution phase there is inhibition of endothelial cell proliferation and migration followed by reconstitution of the basement membrane and recruitment of pericytes and vascular smooth muscle cells. Angiogenesis can take place in both a physiological (in wound healing and female reproductive system) and pathological (tumorigenesis) process, which occurs within the body in response to low oxygen level (hypoxia) and associated stimuli. In order to exploit this process for tissue engineering, the identification of the essential activation signals need to be carried out to enable the recreation of the process *in vitro* or stimulation and acceleration *in vivo*.

5.3   Angiogenic diseases

Although angiogenesis is essential for successful tissue development, as well as grafting and biomaterial-based repair strategies, there are a series of diseases which are affected by either a lack of angiogenesis, or by contrast, uncontrollably high levels of angiogenesis. The main 'angiogenic diseases' are reviewed below, and should be kept in mind when targeting tissue angiogenesis for bone repair.

5.3.1 Cardiovascular disease – myocardial infarction

Myocardial infarction, or heart attack, is the result of clotting in the arterial (sub)occlusion and possible loss of blood flow. This process is thought to originate from haemorrhaging of small blood vessels in the walls of the coronary artery, which in turn are a network of small blood vessels found in atherosclerotic plaques. Endothelial cells are thought to be very proliferative within these plaques, and yet the small vessels are prone to haemorrhage.

Although atherosclerotic plaques are understood to stimulate endothelial cell proliferation, angiogenesis of the damaged arteries may be impaired or insufficient to promote the required repair.[25] In fact, it has been shown in an animal model that older subjects had reduced angiogenesis, and that this was related to lower VEGF expression, although ischaemia has previously been shown to stimulate the angiogenic process.[26] Animal models of vascular and

82   Standardisation in cell and tissue engineering

cardiac ischaemia, using a range of growth factors, gene therapy and endothelial cells and their progenitors, has resulted in promising outcomes.[27-33]

5.3.2   Age-related macular degeneration (AMD)

Also referred to as senile, or diskiform, macular degeneration, age-related macular degeneration (AMD) is an eye condition which results in the loss of eyesight centrally within the visual field. This occurs as a result of damage to the retina, which can occur via one of two mechanisms, depending on whether it is diagnosed as 'dry' or 'wet' AMD. In 'dry' AMD, or geographic atrophy, cellular debris, called 'drusen',[34,35] grow as a hypopigmented spot, which eventually results in vision being obscured. The retina can also detach. The 'wet' form of the disease is thought to be more serious, and is characterised by haemorrhagic fluid arising from a subretinal neovascular membrane, which causes the retinal pigment epithelium (RPE) to detach and disturbs the photoreceptors of the eye. The growth of subretinal blood vessels causes further haemorrhaging and scarring to the eye, and if left untreated, eventually causes degeneration on a faster scale than dry AMD, and can result in legal blindness.[36] There is currently no cure for dry AMD.[37] Treatments include photodynamic therapy, which destroys abnormal vessel growth, as well as taking vitamins and minerals, may also slow down progression of the disease. Angiogenesis-related treatment includes anti-VEGF drugs such as bevacizumab and ranibizumab, which are in fact monoclonal (Fab) antibody fragments, mainly targeted to VEGF-A and blocks the growth factor from promoting angiogenesis in the eye, retarding or reversing the progression of AMD. Use of anti-VEGF drugs has been reported to restore sight completely to some patients, although most will only experience some improvement in their vision.[37]

5.3.3   Cancer – tumour growth

The groundbreaking hypothesis of Folkman *et al.* in 1971[38] described how blood supply and vessel formation, as in angiogenesis, was essential for tumour growth beyond 1–2 mm^3, such that tumour growth is suppressed if kept in a 'prevascular' state.[39,40] The work of Algire[41] and Tannock[42] further promoted the hypothesis, which was unpopular when first proposed, that solid tumours are dependent on angiogenesis. Further experiments suggested that the endothelial cells involved in tumour angiogenesis were releasing angiogenic factors, such as VEGF and FGF[43] that promoted tumour growth, in the presence or absence of endothelial cells.[44,45] Although Folkman's hypothesis was ridiculed at the time, it eventually was accepted and gave rise to a large area of research into anti-angiogenic drugs as a new form of cancer therapy.

In 2004, the first anti-angiogenic drug, bevacuizumab, was approved by the US Food and Drug Administration (FDA) for use in patients.[46,47] This monoclonal anti-VEGF antibody, when used with adjuvant chemotherapy, has been shown to reduce mortality in non-small cell lung cancer.[48] Since then, other drugs targeting VEGF and platelet-derived growth factor (PDGF) via its receptor tyrosine kinase activity, called sorafenib and sunitinib, have also been approved for use by the FDA.[49] In fact, these inhibitors have been shown to be effective on their own for the treatment of metastatic renal-cell cancer. Sorafenib has also been used in the treatment of hepatocellular carcinoma. However, despite the effectiveness of anti-angiogenesis therapy for these cancers, the cost, toxic side-effects and survival benefits have been questioned.[50–52] Many alternatives targeting VEGF for anti-angiogenesis cancer therapy are currently underway,[53] although resistance to anti-angiogenic agents has also proved to be an additional challenge.[54] It may be that for some patients, encouraging angiogenesis in tissue regeneration may have untoward consequences, for example, in cases of tissue reconstruction post-tumour resection and cancer treatment.

5.4 Angiogenesis and bone formation

In bone, microvessels are essential for bone formation, metabolism, healing and remodelling. Both intramembranous and endochondral bone ossification occur in close proximity to vascular ingrowth. In intramembranous ossification, there is an invasion of capillaries that transport bone marrow mesenchymal stem cells (BMSCs, cells derived from bone marrow), which differentiate into osteoblasts and in turn deposit bone matrix. The new vasculature supplies a conduit for the recruitment of cells involved in cartilage resorption and bone deposition.[55] Blood vessels play a crucial role in both phases of bone remodelling. In bone resorption, vessels transport osteoclast precursors to the sites of remodelling. In bone deposition, vessels transport osteoprogenitor cells. Angiogenesis is fundamental to fracture repair. One of the earliest events during bone healing is the reconstruction of intraosseous circulation.[56] Following trauma, disruption of vessels leads to acute hypoxia of the surrounding tissue, as well as to clotting activation. The inflammatory response activates cytokines and GFs that recruit MSCs and ECs to the fracture site. The latter produce PDGF-BB, which contributes to BMSC recruitment.[57] Lack of angiogenesis is considered as a pathogenic cause of non-unions.[58]

5.4.1 Approaches to vascular tissue engineering to encourage bone formation

Engineering complex tissues requires, among other things, a hierarchical vascular network inducing rapid and stable perfusion of the implant; however,

84   Standardisation in cell and tissue engineering

there are current limitations to engineering large tissues. A new, alternative concept is that of prevascularisation, aimed at creating a *de novo* vasculature, whereby a vascular network is formed inside an engineered scaffold graft prior to implantation. This can be achieved by culturing specific cell types in the presence of factors that ensure phenotypic stability and expression, and their interaction with the supporting scaffold.

In vitro vascularisation is a vast and complex topic. This has led to a plethora of methods being proposed in an attempt to either recreate the biological process or trigger it. Implementation of our current knowledge of the various concepts involved in vascularisation has resulted in several strategies. Three elements are needed to engineer an *in vitro* prevascularised construct. These are appropriate material design; angiogenic growth factors; and cell sources.

5.4.2 Scaffold design

Many factors need to be taken into consideration in material design, especially when the construct must support both tissue regeneration and concomitant vascular infiltration. In any tissue structure, cells migrate, proliferate and differentiate in a 3D environment containing a diverse array of biochemical and biophysical signals. Thus, the development of scaffolds to establish *in vitro* vascularisation generally involves specific design criteria to provide 3D structural and logistic templates for tissue development, to control cellular microenvironment and physical-molecular regulatory signals. The architecture in particular has been documented as having a substantial effect upon the rate at which vascularisation takes place *in situ*. In general, the biomaterials used for prevascularised engineered tissues, whether natural or synthetic materials, must consider several key properties:

1. Biocompatibility.
2. Controlled biodegradation – if the resorption rate is too slow, then nascent tissue formation could be impeded, but in contrast if the degradation rate is too fast, the mechanical stability and thus function of engineered tissue could be compromised.[59]
3. Ability to support several cell types of the vasculature (endothelial, osteoblast, fibroblasts and mesenchymal stem cells.
4. Surface property – the material should have ligands binding site and signalling peptide to facilitate cells' surface adhesion.[60]
5. Mechanical property that should mimic the mechanical environment of the target tissue site.[61]
6. Pore structure. Two critical determinants of this are porosity and interconnectivity.

Studies have demonstrated that ingression of host vessels into a scaffold depends upon the size of the pores, with an increase in the rate of

© Woodhead Publishing Limited, 2013

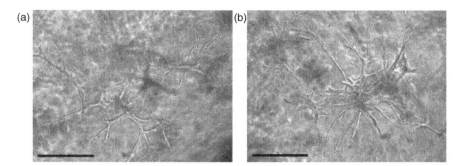

5.2 Tubular formation was observed using phase contrast microscopy in the collagen around CaP scaffolds after 7 days in co-culture system in the presence of VEGF (scale bar = 150 μm).

vascularisation being observed when pores are >250 μm.[24] Nevertheless interconnectivity between these pores is substantially more important, these essential for cell migration, a critical part of the angiogenic process also permeability to the flow transport of nutrients, chemical and metabolic wastes, the absence of an interconnectivity compromises and prevents complete vascularisation of the scaffold even in situations where porosity is high.[62] A wide range of biomaterials have been used to assess prevascularisation; these include natural and synthetic materials such as collagen, fibrin, porous CaP,[63] silk fibroin,[64] polycaprolactone (SPCL),[65] matrigel, PEG and PLGA.[66] Collagen is the most widely used of these materials to prevascularise tissues. It has low antigenicity due to similar molecular structure across different animal species.[67] The degradation, pore size and mechanical properties of collagen can be manipulated by using different pH or temperatures during polymerisation. Collagen cross-linking is temperature-dependent, and therefore a rapid polymerisation can be carried out at room temperature or higher. In addition, collagen can synergise with other materials to promote angiogenesis. For example, in our lab type I rat tail collagen was applied with porous calcium phosphate scaffold seeded with human umbilical vein endothelial cells (HUVECs) and human osteoblasts (HOBs). This construct gave rise to a rudimentary tubular-like network *in vitro* in 7 days in a conditioned medium supplemented with exogenous VEGF growth factor (Fig. 5.2).

Fibrin is a naturally occurring material which has an important role in the final steps of the coagulation cascade in wound healing,[68] and has been used previously as an effective tissue engineering scaffold.[69–71] Fibrin is formed in the body by the enzymatic cleavage of fibrinogen by thrombin. Fibrin then undergoes a polymerisation reaction with other fibrin molecules, as well as cross-linking.[72] Cells naturally express urokinase plasminogen activator and tissue plasminogen activator, serine proteases that activate the fibrin-degradation enzyme plasmin,[73] and this process, in addition to the action of macrophages and multi-nucleated foreign body cells, naturally degrades fibrin over time. Fibrin glue has been used for many years as a surgical haemostatic agent,[74,75] as well as other surgical uses, for example, preventing air leaks after pulmonary resection.[76] The availability of fibrin and its simple application system make it easy to obtain and use. Combinations of fibrin and nonresorbable HA have been used throughout Europe for over a decade in the reconstruction of maxillofacial and dental defects.[77] The fibrin phase has proved effective in moulding the ceramic scaffolds and holding the granules/powder in place as natural mineral deposition occurs. Fibrin has been shown to increase the level of the angiogenic factor VEGF, which could enhance the healing effect of fibrin scaffolds, as reported, for example, in skin regeneration research.[78] Fibrin may also promote neovascularisation to enhance healing and bone formation and organisation.[68] A study by Abiraman *et al.* demonstrated that the ectopic implants coated in fibrin glue had a noticeable neovascularisation and evidence of bone formation, compared to controls.[79]

Matrigel® is the commercial name for ECM extracted from Engelbreth-Holm-Swarm mouse sarcoma cells. It is one of the most abundantly used matrices for the study of angiogenesis. The main components are collagen type IV, laminin, heparin sulphate proteoglycan and entactin.[80] Matrigel has been useful for studying the mechanisms of neovascularisation and angiogenesis. For example, a recent study by our group demonstrated that the optimal condition for culture of HUVECs in the presence of 10 ng/mL VEGF allowed endothelial cells to form a tubular branching morphology in matrigel within 48 h (Fig. 5.3). Although matrigel is widely utilised as a 3D scaffold material for *in vitro* studies of vasculature network, major disadvantages of matrigel are batch-to-batch variability and its sarcoma cell origin which makes it unfavourable for clinical translational applications of vascularised material.

5.4.3 Angiogenic growth factors

Currently, several strategies exist for vascularisation of tissue-engineered biological or synthetic scaffolds. One approach is to integrate or functionalise the growth factors to the scaffold. It is well established that the addition of angiogenic factors can stimulate different stages of blood vessel formation and hence enhance vascularisation of tissue-engineered construct after implantation. Growth factors that are involved in the vascularisation process can be categorised according to their action. Almost 30 factors influencing angiogenesis have been described in the literature.[16,81] Two main categories exist: (i) those that directly affect endothelial cell activity by stimulating the conversion of quiescent cells to the angiogenic phenotype, thus controlling initial formation of new vessels such as vascular endothelial growth factor

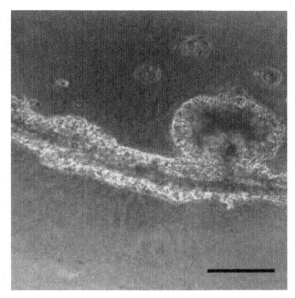

5.3 HUVECs grown on Matrigel® in the presence of VEGF (10 ng/mL), cell aggregation and lumen or capillary-like structure formation was observed at 48 h (scale bar = 200 μm).

(VEGF), hepatocyte growth factor (HGF) and basic fibroblastic growth factor (bFGF); and (ii) those that do not directly affect endothelial activity but instead control maturation of these vessels, such as platelet-derived growth factor (PDGF), transforming growth factor β (TGF-β) and angiopoeitin 1 (Ang1).[82] A summary of these angiogenic growth factors can be found in Table 5.1. Newly formed vessels are often 'disorganised, leaky and haemorrhagic'; maturation (also known as stabilisation) is therefore essential. Stabilisation involves the recruitment and incorporation of mural cells, such as pericytes and smooth muscle cells, to provide support, which ultimately yields a new functional vascular network.[83] Despite the importance of maturation in vessel integrity, present work revolves around the first group of pro-angiogenic factors. These factors have been identified as being imperative in stimulating quiescent endothelial cells *in vivo* and initiating immature vessel formation *in vitro*.[84]

5.4.4 Vascular endothelial growth factor (VEGF)

Most research to improve the biological properties of bone tissue engineering has focused on the incorporation of osteoinduction growth factors such as bone morphogenetic proteins (BMPs). More recently, angiogenic growth

88 Standardisation in cell and tissue engineering

Table 5.1 Summary of angiogenic growth factors, organised by target cell population and use in either bone formation or remodelling

	Vascular formation	Bone formation	Bone remodelling
Osteoblast	VEGF, FGF, IGF, PDGF and TGF-β		
Endothelial		PDGF-AB, TGF-β, FGF, EGF, OPG, BMP-2 ECs activate OB to express ALP, increase collagen type I	RANKL, OPG, IL-6, PDGF

BMP - bone morphogenetic proteins; EGF - epidermal growth factor; FGF - fibroblast growth factor; IGF - insulin-like growth factor; OPG - osteoprotegerin; PDGF - platelet-derived growth factor; RANKL - receptor activator of nuclear factor kappa-B ligand; TGF-β- transforming growth factor beta; VEGF - vascular endothelial growth factor.

factors such as vascular endothelial growth factor (VEGF) have gained increasing attention due to its pivotal contribution to angiogenesis during bone healing. Several studies provided evidence that local delivery of VEGF from the carrier biomaterials improved the healing of critical size bone defects.[85,86] VEGF is a major regulator of neovascularisation, which exerts its effect via receptors upon the endothelial cell. It is regarded as the most potent mitogen for micro- and macrovascular endothelial cells. However, VEGF has been discovered to possess a vast array of effects, which include stimulating endothelial cell recruitment, differentiation, survival, microvascular permeability and the promotion of neovascularisation within three-dimensional *in vitro* models.[87] The importance of VEGF as a vital mediator both *in vitro* and *in vivo* has been well documented. Gerber *et al.* have shown that VEGF prevents apoptosis by phosphatidylinositol (PI)-3-kinase-Akt pathway and also induces expression of anti-apoptotic proteins in endothelial cells.[88] Using *in vivo* knockout mice models, low concentrations (VEGF$^{+/-}$) or no (VEGF$^{-/-}$) expression was found to be lethal.[89] VEGF mediates its ability to enhance survival through the up-regulation of the expression of anti-apoptotic proteins, which inhibit apoptosis that is induced by serum starvation. VEGF is produced as five homodimeric isoforms, which differ in their expression levels and in their localisation. VEGF$_{121}$ and VEGF$_{165}$ are the most abundant isoforms. VEGF$_{121}$ is an acidic soluble polypeptide that does not bind to heparin sulfate proteoglycans on the cell surface. VEGF$_{165}$ has intermediary properties, as it is secreted but a significant fraction remains bound to the cell surface and ECM.[90] As with any physiological process the correct level of stimulation is important. The dose-dependent effect of VEGF is a typical example of this; too low a dose fails to provide enough stimulation for the cells. However beyond the optimum dose, high doses cause vessel leakage through increased permeability and induce the formation of malformed, non-functional blood vessels.[91] The release kinetic of VEGF appears to be an important factor to enhance biomaterial vascularisation

© Woodhead Publishing Limited, 2013

Vascularisation of tissue-engineered constructs   89

and bone formation. Sustained release of VEGF increased the efficacy of VEGF delivery, demonstrating that a prolonged bioavailability of low concentrations of VEGF is beneficial for bone regeneration.[92]

VEGF has predominantly been considered endothelial cell-specific; however, previous reports have confirmed the presence of VEGF receptors, flt-1 and/or KDR on numerous other cell types, including osteoblasts.[93]

Co-culture models aim to simulate as closely as possible the *in vivo* situation and allow the interaction of different cell types to be examined. It has been shown that osteoblasts release VEGF in response to a number of stimuli and express receptors for VEGF in a differentiation-dependent manner, thus indicating that cells react to paracrine factors released by neighbouring cells, influencing cell function and gene expression. Our group has shown an increase in tubule formation *in vitro*, in co-cultures where osteoblasts were in co-existence with other cell types, and where the production of extracellular matrix and stimulatory factors such as VEGF, essentially provided a pro-angiogenic effect on the endothelial cells (unpublished data).

5.4.5 β-Fibroblast growth factor (βFGF)

Fibroblast growth factors comprise two main family members: acidic FGF (αFGF) which is mainly expressed in neuronal tissues, bone and tumours; and basic fibroblast growth factor (βFGF or FGF-2) which is widely distributed in both normal and malignant tissues. βFGF acts as a strong fibroblast mitogen and was identified as the first EC mitogen. βFGF is also a chemotactic factor for EC and is highly angiogenic. As can be seen, the actions of VEGF and FGF are similar. However, FGF is not specific to EC as FGF receptors are also expressed in smooth muscle cells, fibroblasts and myoblasts.[94] FGF and VEGF exert their effects via specific binding to cell surface-expressed receptors equipped with tyrosine kinase activity. Activation of the receptor kinase activity allows coupling to downstream signal transduction pathways that regulate proliferation, migration and differentiation of EC.[95] Hepatocyte growth factor (HGF) is also a pleiotropic factor that participates in EC migration and proliferation. HGF has been described as a potent mitogen and induces the formation of new blood vessels *in vivo*.[96]

5.4.6 Transforming growth factor-β (TGF-β), platelet-derived growth factor (PDGF) and angiopoeitin (Ang)

The roles of TGF-β, PDGF and Ang within the domain of vascularisation are rather complex. They are often referred to as indirect angiogenic factors

90   Standardisation in cell and tissue engineering

as they are considered to exert their effects not directly upon endothelial cells but instead through intermediate cells.

TGF-β is an important regulator of tissue morphogenesis and a potent inhibitor of proliferation for most cell types.[97] Approximately 50% of mice genetically deficient in TGF-β1 die *in utero* and show defective vasculogenesis.[98] TGF-β1 has multiple effects on vascular endothelial cells. *In vivo* TGF-β1 induces angiogenesis.[99–101] However, *in vitro* it inhibits endothelial cell proliferation, migration and proteolytic activity.[102] Notably, TGF-β1 induces endothelial cell apoptosis, opposing the pro-survival activity of VEGF. During angiogenesis apoptosis is required for pruning the forming vascular network, and inhibition of apoptosis results in formation of abnormal vessels. In addition, apoptosis controls cell functions required for capillary morphogenesis *in vitro* and *in vivo*.[103] Because of its inhibitory effects on endothelial cells it has been proposed that TGF-β1 induces angiogenesis *in vivo* through an indirect mechanism, by inducing expression of VEGF and/or other angiogenic factors in epithelial or other cell types.[104] However, several observations indicate that TGF-β1 has important direct effects on angiogenesis.

PDGF is a known potent mediator of growth and motility of fibroblasts and smooth muscle cells. PDGF receptors have been identified on a subset of endothelial cells (within developing endothelial tubes or microvascular endothelial cells). PDGF-BB could act directly on angiogenic endothelial cells via PDGF B-receptors to change cell shape, to modulate the generation and degradation of extracellular matrix, to regulate the generation of traction forces, to alter the proteolytic potential, and to modify the response of angiogenic endothelial cells to other growth regulatory molecules.[105]

Angiopoietins have been described as ligands for EC receptor 'Tie-2'. Only angiopoietin-1 (Ang-1) promotes endothelial cell survival, sprouting and tube formation whilst Ang-2 is a competitive inhibitor of Ang-1. The proposed function of Ang-2 is that it serves to inhibit the interaction between EC and the supporting cells in order to facilitate EC migration from vessels to form new capillaries. Ang is needed as a vascular stabilizing factor organising and limiting the angiogenesis response and protecting from pathological consequences, such as tissue fibrosis.[106]

5.4.7   Other stimulants for angiogenesis – hypoxia

Hypoxia, or lowering of oxygen tension below normal levels, stimulates the natural angiogenic response. Collateral blood vessels develop, for example, in ischaemic tissues[107] via an oxygen-sensing heme protein that induces hypoxia-inducible factor-1 and -2 alpha (HIF-1α and HIF-2α).[108] The alpha subunit is normally degraded under normoxic conditions (normal oxygen tension), but its degradation is reduced in hypoxic conditions. Hypoxia

© Woodhead Publishing Limited, 2013

has also been shown to stabilise the VEGF mRNA transcript, protecting it from degradation.[109] With regard to bone formation, osteoblastic activation of the HIF signalling pathway was shown to prevent ovariectomy-induced bone loss in mice and increase blood vessel formation. This suggests that targeting hypoxia-induced pathways may also help prevent oestrogen deficiency-related bone loss.[110]

5.5 Cell sources for vascular tissue engineering

In general, cells for an engineered tissue are strictly isolated from the patient's own body to minimise any risk of immunorejection. Several types of cell are potential sources for vascular tissue engineering, including mature ECs, HUVEC, endothelial progenitor cells (EPCs) and endothelial cell lines. Most mature ECs have demonstrated their ability to form capillary-like structures; however, their rapid *in vitro* senescence limits their utility for cell-based therapies. To address this limit, attempts have been made to immortalise EC lines by genetic manipulation, resulting in an unlimited proliferation. Nevertheless, these genetically modified cells might not provide a reliable source of cells, due to the unforeseen risk of tumour development if proliferation is not carefully controlled.

Despite the use of adult mature ECs in tissue engineering of vascularisation, problems of decline in proliferation and differentiation abilities of these cells have been reported. Another potential source is embryonic stem cell (ESC)-derived ECs. These pluripotent ESCs are characterised by their ability to proliferate in an unlimited manner, capable of multiple passages while maintaining normal karyotype and function, making them an alternative source of cells for vascular regeneration. Levenberg *et al.* first described directing hESCs towards an endothelial fate[111] by culturing hESCs in conditions tuned for the formation of ECs, then examining them for endothelial-specific markers to confirm the existence of ECs including CD31 or PECAM1, VE-cadherin and VWF. The two main methods of purifying ECs from hESCs are (i) culturing in endothelial maturation conditions by supplementing either the feeder layer (e.g., matrigel) or the media; and (ii) selecting cells derived from hESC displaying endothelial-specific markers.[112]

5.6 Co-culture of cells: the interactions between angiogenesis and osteogenesis

Much of the current *in vitro* work focuses upon monoculture of cells. This approach is somewhat flawed in that it provides a poor representation of the *in vivo* scenario. Within the body, cells are constantly exposed to a wide variety of biochemical signals, and they engage in complex interactions with

92   Standardisation in cell and tissue engineering

other cell types that provide signal cues for stability and survival. These consist of several co-existent cell types via both direct (cell–cell contact) and indirect (e.g., growth factor-induced) signalling. The optimum physiological dose and timing of growth signalling factors is important, as is the effect of exogenous growth factors released by neighbouring cells. Thus, contact with other cell types through co-culture would be a necessary pre-requisite to mimic natural cell niche and to engineer complex tissue (see Plate III in the colour section between pages 134 and 135). ECs produce GFs which contribute to the recruitment of stem cells and act as cues for osteoblast differentiation. Angiogenic GFs are also involved in bone formation; for example, VEGF plays an important role in endochondral ossification, where its functions are mediated by cbfa-1/runx-2.[113] VEGF production is increased by BMP-2, -4, and -6,[93] and by TGF-β1.[114] FGF-2 also stimulates the proliferation and differentiation of osteoblasts[115] and accelerates fracture repair when added to the early healing stage. Table 5.1 compares cell co-culture systems, using different cell combinations, and describes their relative VEGF production and angiogenic/osteogenic efficacy.

5.7   Strategies to induce *in vitro* prevascularisation

Angiogenic factors are initiators of neovascularisation by activating endothelial cells and promoting assembly of cells and subsequent vessel formation. These factors, whether applied exogenously or released in a paracrine manner, work synchronously with cytokines such as PDGF to regenerate endothelial tubule formation. Prevascularisation of grafts with a capillary network aims to accelerate anastomosis with host tissue.

To evaluate the potential of bone scaffolds to be prevasularised, the 3D construct must support different cell types to proliferate, differentiate and allow ECs to develop a functional vasculature to supply the cells with oxygen and nutrients. Angiogenesis is a complex process brought about by sequential cell–cell and cell–matrix interactions. Furthermore, cells can act reciprocally through direct physical contact, releasing specific factors and promoting ECM production. However, promoting angiogenesis is a challenging task. Our group has optimised a co-culture model to mimic the natural environmental niche comprising HUVECS, HOBs and MSCs. HUVECs are traditionally used as the favoured cell type for *in vitro* vascular studies due to ease of isolation and culture. In cell models where different cell types have to co-exist, culture conditions need to be optimised for all cell types to ensure their survival and potential to differentiate and maintain their phenotype. Furthermore, each cell type has different growth rates and nutritional requirements, which should be considered when developing appropriate angiogenic models. The conventional strategy for cell-seeded scaffolds is to expand the cells in culture and then seed them on the test scaffolds. When combining different cell

© Woodhead Publishing Limited, 2013

types, the order of cell seeding and the choice of cells can have important consequences. Both ECs and HOBs may be seeded at the same time; however, there is an advantage to seeding HOBs before ECs, as this results in the production of an ECM which provides a favourable niche for the formation of tubules. ECs have a faster growth rate than HOBs, and hence rapidly colonise the scaffold. A similar concept, previously reported by Kirkpatrick *et al.* (2011), showed that pre-culturing with osteoblasts *in vitro* before transplantation *in vivo* could induce a more homogeneous vascularisation of the scaffold.[64] To date, endothelial cells have been studied in static *in vitro* cultures; more recently, however, the use of dynamic conditions has been investigated. Matsuda *et al.* (2004) have demonstrated that static seeded endothelial-lined vascular grafts showed a high incidence of cell loss, at levels equivalent to 24 h arterial circulation sheer stress.[116] In optimising the cell niche, our group investigated the combination of ratios of different cell types, in order to achieve an ideal cell number of each type with the capacity to still be able to express their phenotypic characteristics. Our investigations confirmed that the optimal HUVEC: HOB ratio was 60:40 (Fig. 5.4a), and HUVEC:MSC was 30:70 (Fig. 5.4b). Immunofluorescent cell tracking was used to confirm the equal growth of each cell type, cell counts using fluorescent microscopy provided quantitative evidence of the presence of both cell types for the time periods studied and both cell types showed good viability (see Plate IV in the colour section between pages 134 and 135).

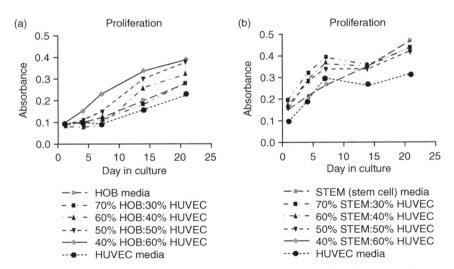

5.4 Cell proliferation observed over a 28-day period to determine the ideal medium for maximum cell growth within co-culture comprising (a) HUVECs + HOBs and (b) HUVECs + MSCs. STEM, stem cell media: Dulbecco's Modified Eagles's Medium (Sigma D6429) supplemented with 20% fetal calf serum, 100 units/ml penicillin and 0.1 mg/ml streptomycin.

94   Standardisation in cell and tissue engineering

5.8   Tubular formation

The successful vascularisation of scaffolds is dependent on the formation of a tissue-like self-assembly of extensive microcapillary-like structures. This endothelial phenotype outgrowth of EC inside the scaffold structure can be demonstrated *in vitro* and has been shown by several methods, such as scanning electron microscopy (SEM), or immunofluorescent staining for endothelial markers such as CD31, VE-cadherin and VWF. We observed EC invasion and lumenal morphogenesis in 3D collagen matrices that mimicked angiogenic sprouting (Fig. 5.5a). Viewing with SEM demonstrated the presence of the tubule structures within a calcium phosphate scaffold (Fig. 5.5b). The presence of vessel networks formed by ECs within scaffold were identified in *in vitro* tissue stained with anti-human CD31 antibody (Fig. 5.6).

5.9   Conclusion

The success of any tissue-engineered implant is largely dependent on the ability to provide oxygen to the cells involved. In both pre-seeded scaffolds and scaffolds where the cell–material interaction is to be induced *in vivo*, successful vascularisation is essential for survival and integration of the implant graft or scaffold.

Angiogenesis is dependent upon recruitment of quiescent stem cells and progenitor cells; the bioavailability of specific cytokines and growth factors is paramount to facilitate this recruitment. VEGF is the most potent and specific of the growth factors that regulate angiogenesis. Tissue regeneration will only proceed *in vivo* if vascularisation from the host tissue to the implanted biomaterial is established, which will only occur if the microvasculature established within the biomaterial or graft is able to 'hook up' to the existing vasculature to re-establish blood flow.

To address the need for vascularisation numerous strategies have been described which require sophisticated scaffolds and the presence of angiogenic factors to enhance ingrowth of vessels from the host. The addition of specific cells able to generate a plexus-like vascular network that could perfuse the scaffold or graft upon anastomosis could potentially give rise to significantly superior quality grafts. This chapter has described how it is possible to model *in vitro* the physiological environmental niche in order to understand the complex cell–cell and cell–material interactions and how vascularisation can be influenced by cell types, growth factors and other factors. Vascularisation also plays a role in a number of diseases, some of which have been discussed in this chapter. Understanding the process of angiogenesis and vascularisation in disease mechanisms has led to the development of a range of new therapies.

The development of large three-dimensional engineered tissue constructs requires convective transport of nutrients to the inner part of the tissue quickly after implantation. Incorporation of ECs in tissue constructs may

© Woodhead Publishing Limited, 2013

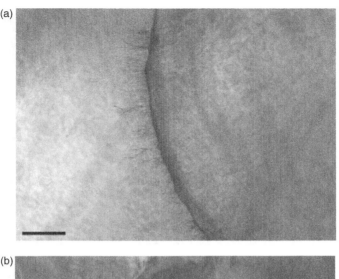

5.5 (a) Tubular formation observed using phase contrast microscopy in the collagen around scaffolds after 7 days in co-culture system in the presence of VEGF (scale bar = 200 μm). (b) SEM revealed tubule formation even within the inner recesses of the calcium phosphate scaffold.

provide an effective method to develop vessel networks, thus contributing to the successful integration of the graft with the host tissue.

Ongoing studies have begun to demonstrate the feasibility of the ultimate goal, to generate vascularised grafts that maintain anatomically desirable shape and dimensions and successfully integrate with the host tissue. However, challenges still remain to be addressed before such systems can be used clinically.

96 Standardisation in cell and tissue engineering

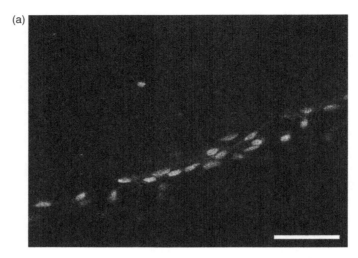

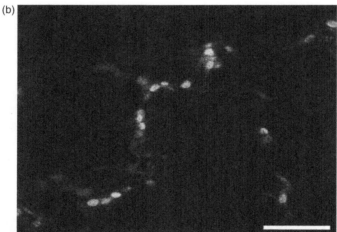

5.6 Three-dimensional interconnected networks formed within 7 days by co-coture of HUVEC/HOB seeded on calcium phosphate porous scaffold, fluorescent micrographs of tubular structures stained with (a) CD31 and (b) vWF antibodies.

5.10 References

1. Rouwkema, J., Rivron, N.C. and van Blitterswijk, C.A. Vascularization in tissue engineering. *Trends Biotechnol* **26**, 434–441 (2008).
2. Janssen, F.W., Oostra, J., Oorschot, A.V. and van Blitterswijk, C.A. A perfusion bioreactor system capable of producing clinically relevant volumes of tissue-engineered bone: In vivo bone formation showing proof of concept. *Biomaterials* **27**, 315–323 (2006).

Vascularisation of tissue-engineered constructs 97

3. Clark, E.R. and Clark, E.L. Microscopic observations on the growth of blood capillaries in the living mammal. *Am J Anat* **64**, 251–301 (1939).

4. Tian, L. and George, S. Biomaterials to prevascularize engineered tissues. *J Cardiovasc Transl Res* **4**, 685–698 (2011).

5. Evans, H.M. On the development of the aortae, cardinal and umbilical veins, and the other blood vessels of vertebrate embryos from capillaries. *Anat Rec* **3**, 498–519 (1909).

6. Sabin, F.R. Origin and development of the primitive vessels of the chick and of the pig. *Carnegie Contrib Embryol* **9**, 215–262 (1917).

7. Sabin, F.R. Studies on the origin of the blood vessels and of red blood corpuscles as seen in the living blastoderm of chick during the second day of incubation. *Carnegie Contrib Embryol* **9**, 215–262 (1920).

8. Hughes, A.F.W. The histogenesis of the arteries in the chick embryo. *J Anat* **77**, 266–287 (1942).

9. Haar, J.L. and Ackerman, G.A. A phase and electron microscopic study of vasculogenesis and erythropoiesis in the yolk sac of the mouse. *Anat Rec* **170**, 199–223 (1971).

10. Wilting, J. and Christ, B. Embryonic angiogenesis: a review. *Naturwissenschaften* **83**, 153–164 (1996).

11. Slack, J.M., Darlington, B.G., Heath, J.K. and Godsave, S.F. Mesoderm induction in early Xenopus embryos by heparin-binding growth factors. *Nature* **326**, 197–200 (1987).

12. Kimelman, D. and Kirschner, M. Synergistic induction of mesoderm by FGF and TGF-beta and the identification of an mRNA coding for FGF in the early Xenopus embryo. *Cell* **51**, 869–877 (1987).

13. Woodland, H.R. Mesoderm formation in Xenopus. *Cell* **59**, 767–770 (1989).

14. His, W. Lecithoblast und Angioblast der Wirbeltiere. *Abhandl Math-phys Ges Wiss* **26**, 171–328 (1900).

15. Benninghoff, A., Hartmann, A. and Hellmann, T. (eds) *Handbuch Der Mikroskopischen Anatomie Des Menschen* (Springer, Berlin, 1930).

16. Risau, W. Embryonic angiogenesis factors. *Pharmacol Ther* **51**, 371–376 (1991).

17. Ausprunk, D.H. and Folkman, J. Migration and proliferation of endothelial cells in preformed and newly formed blood vessels during tumor angiogenesis. *Microvasc Res* **14**, 53–65 (1977).

18. Cogan, D.G. Vascularization of the cornea; ats experimental induction by small lesions and a new theory of its pathogenesis. *Arch Ophthal* **41**, 406–416 (1949).

19. Wilting, J., Christ, B., Bokeloh, M. and Weich, H.A. In vivo effects of vascular endothelial growth factor on the chicken chorioallantoic membrane. *Cell Tissue Res* **274**, 163–172 (1993).

20. Rizzo, V., Steinfeld, R., Kyriakides, C. and DeFouw, D.O. The microvascular unit of the 6-day chick chorioallantoic membrane: a fluorescent confocal microscopic and ultrastructural morphometric analysis of endothelial permselectivity. *Microvasc Res* **46**, 320–332 (1993).

21. Rizzo, V., Kim, D., Duran, W.N. and DeFouw, D.O. Differentiation of the microvascular endothelium during early angiogenesis and respiratory onset in the chick chorioallantoic membrane. *Tissue Cell* **27**, 159–166 (1995).

22. Rizzo, V., Kim, D., Duran, W.N. and DeFouw, D.O. Ontogeny of microvascular permeability to macromolecules in the chick chorioallantoic membrane during normal angiogenesis. *Microvasc Res* **49**, 49–63 (1995).

© Woodhead Publishing Limited, 2013

98 Standardisation in cell and tissue engineering

23. Asahara, T. , Murohara, T., Sullivan, A., Silver, M., van der Zee, R., Li, T., Witzenbichler, B., Schatteman, G. and Isner, J.M. Isolation of Putative Progenitor Endothelial Cells for Angiogenesis. *Science* **275**, 964–966 (1997).
24. Drake, C.J., Hungerford, J.E. and Little, C.D. Morphogenesis of the first blood vessels. *Ann NY Acad Sci* **857**, 155–179 (1998).
25. Freedman, S.B. and Isner, J.M. Therapeutic angiogenesis for ischemic cardio-vascular disease. *J Mol Cell Cardiol* **33**, 379–393 (2001).
26. Rivard, A. Fabre, J.-E., Silver, M., Chen, D., Murohara, T., Kearney, M., Magner, M., Asahara, T. and Isner, J.M. Age-dependent impairment of angiogenesis. *Circulation* **99**, 111–120 (1999).
27. Unger, E.F., Banai, S., Shou, M., Lazarous, D.F., Jaklitsch, M.T., Scheinowitz, M., Correa, R., Klingbeil, C. and Epstein, S.E. Basic fibroblast growth factor enhances myocardial collateral flow in a canine model. *Am J Physiol* **266**, H1588–H1595 (1994).
28. Lazarous, D.F., Scheinowitz, M., Shou, M., Hodge, E., Sharmini Rajanayagam, M.A., Hunsberger, S., Robison Jr, W.G., Stiber, J.A., Correa, R., Epstein, S.E. and E.F. Unger Effects of chronic systemic administration of basic fibroblast growth factor on collateral development in the canine heart. *Circulation* **91**, 145–153 (1995).
29. Rajanayagam, M.A., Shou, M., Thirumurti, V., Lazarous, D.F., Quyyumi, A.A., Goncalves, L., Stiber, J., Epstein, S.E. and Unger, E.F. Intracoronary basic fibro-blast growth factor enhances myocardial collateral perfusion in dogs. *J Am Coll Cardiol* **35**, 519–526 (2000).
30. Vale, P.R., Losordo, D.W., Tkebuchava, T., Chen, D., Milliken, C.E. and Isner, J.M. Catheter-based myocardial gene transfer utilizing nonfluoroscopic elec-tromechanical left ventricular mapping. *J Am Coll Cardiol* **34**, 246–254 (1999).
31. Agudelo, C.A., Tachibana, Y., Hurtado, A.F., Ose, T., Iida, H. and Yamaoka, T. The use of magnetic resonance cell tracking to monitor endothelial progenitor cells in a rat hindlimb ischemic model. *Biomaterials* **33**, 2439–2348 (2012).
32. Fierro, F.A., Kalomoiris, S., Sondergaard, C.S. and Nolta, J.A. Effects on pro-liferation and differentiation of multipotent bone marrow stromal cells engi-neered to express growth factors for combined cell and gene therapy. *Stem Cells* **29**, 1727–1737 (2011).
33. Kuliszewski, M.A., Kobulnik, J., Lindner, J.R., Stewart, D.J. and Leong-Poi, H. Vascular gene transfer of SDF-1 promotes endothelial progenitor cell engraftment and enhances angiogenesis in ischemic muscle. *Mol Ther* **19**, 895–902 (2011).
34. Abdelsalam, A., Del Priore, L. and Zarbin, M.A. Drusen in age-related macular degeneration: pathogenesis, natural course, and laser photocoagulation-induced regression. *Surv Ophthalmol* **44**, 1–29 (1999).
35. Donders, F.C. Beitraege zur pathologischen Anatomie des Auges. *Arch Ophthamol* **1**, 106–118. (1854).
36. de Jong, P.T. Age-related macular degeneration. *N Engl J Med* **355**, 1474–1485 (2006).
37. NHS Choices: Macular degeneration. http://www.nhs.uk/Conditions/Macular-degeneration/Pages/Introduction.aspx.
38. Folkman, J. Tumor angiogenesis: therapeutic implications. *N Engl J Med* **285**, 1182–1186 (1971).

Vascularisation of tissue-engineered constructs 99

39. Folkman, J. How is blood vessel growth regulated in normal and neoplastic tissue? G.H.A. Clowes Memorial Award lecture. *Cancer Res* **46**, 467–73 (1986).
40. Kerbel, R.S. Tumor angiogenesis. *N Engl J Med* **358**, 2039–2049 (2008).
41. Algire, G.H., Chalkley, H.W., Legallais, F.Y. and Park, H.D. Vascular reactions of normal and malignant tumors in vivo. I. Vascular reactions of mice to wounds and to normal and neoplastic transplants. *J Natl Cancer Inst* **6**, 73–85 (1945).
42. Tannock, I.F. The relation between cell proliferation and the vascular system in a transplanted mouse mammary tumour. *Br J Cancer* **22**, 258–273 (1968).
43. Folkman, J. and Shing, Y. Control of angiogenesis by heparin and other sulfated polysaccharides. *Adv Exp Med Biol* **313**, 355–364 (1992).
44. Ferrara, N. VEGF and the quest for tumour angiogenesis factors. *Nat Rev Cancer* **2**, 795–803 (2002).
45. Hicklin, D.J. and Ellis, L.M. Role of the vascular endothelial growth factor pathway in tumor growth and angiogenesis. *J Clin Oncol* **23**, 1011–1027 (2005).
46. Ferrara, N., Hillan, K.J., Gerber, H.P. and Novotny, W. Discovery and development of bevacizumab, an anti-VEGF antibody for treating cancer. *Nat Rev Drug Discov* **3**, 391–400 (2004).
47. Folkman, J. Angiogenesis: an organizing principle for drug discovery? *Nat Rev Drug Discov* **6**, 273–286 (2007).
48. Hurwitz, H., Fehrenbacher, L., Novotny, W., Cartwright, T., Hainsworth, J., Heim, W., Berlin, J., Baron, A., Griffing, S., Holmgren, E., Ferrara, N., Fyfe, G., Rogers, B., Ross, R. and Kabbinavar, F. Bevacizumab plus irinotecan, fluorouracil, and leucovorin for metastatic colorectal cancer. *N Engl J Med* **350**, 2335–2342 (2004).
49. Faivre, S., Demetri, G., Sargent, W. and Raymond, E. Molecular basis for sunitinib efficacy and future clinical development. *Nat Rev Drug Discov* **6**, 734–745 (2007).
50. Berenson, A. A cancer drug shows promise, at a price that many can't pay. *The New York Times* (2006). http://www.nytimes.com/2006/02/15/business/15drug.html
51. Eskens, F.A. and Verweij, J. The clinical toxicity profile of vascular endothelial growth factor (VEGF) and vascular endothelial growth factor receptor (VEGFR) targeting angiogenesis inhibitors; a review. *Eur J Cancer* **42**, 3127–3139 (2006).
52. Verheul, H.M. and Pinedo, H.M. Possible molecular mechanisms involved in the toxicity of angiogenesis inhibition. *Nat Rev Cancer* **7**, 475–485 (2007).
53. Pan, Q., Chanthery, Y., Liang, W.C., Stawicki, S., Mak, J., Rathore, N., Tong, R.K., Kowalski, J., Yee, S.F., Pacheco, G., Ross, S., Cheng, Z., Le Couter, J., Plowman, G., Peale, F., Koch, A.W., Wu, Y., Bagri, A., Tessier-Lavigne, M. and Watts, R.J. Blocking neuropilin-1 function has an additive effect with anti-VEGF to inhibit tumor growth. *Cancer Cell* **11**, 53–67 (2007).
54. Miller, K.D., Sweeney, C.J. and Sledge, G.W., Jr. The Snark is a Boojum: the continuing problem of drug resistance in the antiangiogenic era. *Ann Oncol* **14**, 20–28 (2003).
55. Klagsbrun, J.H.M. Cartilage to bone—Angiogenesis leads the way. *Nat Med* **5**, 617–618 (1999).
56. Schindeler, A., McDonald, M.M., Bokko, P. and Little, D.G. Bone remodeling during fracture repair: the cellular picture. *Semin Cell Dev Biol* **19**, 459–466 (2008).

100 Standardisation in cell and tissue engineering

57. Cenni, E., Ciapetti, G., Granchi, D., Fotia, C., Perut, F., Giunti, A. and Baldini, N. Endothelial cells incubated with platelet-rich plasma express PDGF-B and ICAM-1 and induce bone marrow stromal cell migration. *J Orthopaed Res* **27**, 1493–1498 (2009).

58. Fang, T.D., Salim, A., Xia, W., Nacamuli, R.P., Guccione, S., Song, H.M., Carano, R.A., Filvaroff, E.H., Bednarski, M.D., Giaccia, A.J. and Longaker, M.T. Angiogenesis is required for successful bone induction during distraction osteogenesis. *J Bone Miner Res* **20**, 1114–1124 (2005).

59. Sill, T.J. and von Recum, H.A. Electrospinning: applications in drug delivery and tissue engineering. *Biomaterials* **29**, 1989–2006 (2008).

60. Chan, B. and Leong, K. Scaffolding in tissue engineering: general approaches and tissue-specific considerations. *Eur Spine J* **17**, 467–479 (2008).

61. Geckil, H., Xu, F., Zhang, X., Moon, S. and Demirci, U. Engineering hydrogels as extracellular matrix mimics. *Nanomedicine* **5**, 469–484 (2010).

62. Karageorgiou, V. and Kaplan, D. Porosity of 3D biomaterial scaffolds and osteogenesis. *Biomaterials* **26**, 5474–5491 (2005).

63. Unger, R.E., Ghanaati, S., Orth, C., Sartoris, A., Barbeck, M., Halstenberg, S., Motta, A., Migliaresi, C. and Kirkpatrick, C.J. The rapid anastomosis between prevascularized networks on silk fibroin scaffolds generated in vitro with cocultures of human microvascular endothelial and osteoblast cells and the host vasculature. *Biomaterials* **31**, 6959–6967 (2010).

64. Ghanaati, S., Unger, R.E., Webber, M.J., Barbeck, M., Orth, C., Kirkpatrick, J.A., Booms, P., Motta, A., Migliaresi, C., Sader, R.A. and Kirkpatrick, C.J. Scaffold vascularization in vivo driven by primary human osteoblasts in concert with host inflammatory cells. *Biomaterials* **32**, 8150–8160 (2011).

65. Hofmann, A., Ritz, U., Verrier, S., Eglin, D., Alini, M., Fuchs, S., Kirkpatrick, C.J. and Rommens, P.M. The effect of human osteoblasts on proliferation and neo-vessel formation of human umbilical vein endothelial cells in a long-term 3D co-culture on polyurethane scaffolds. *Biomaterials* **29**, 4217–4226 (2008).

66. Fuchs, S., Ghanaati, S., Orth, C., Barbeck, M., Kolbe, M., Hofmann, A., Eblenkamp, M., Gomes, M., Reis, R.L. and Kirkpatrick, C.J. Contribution of outgrowth endothelial cells from human peripheral blood on *in vivo* vascularization of bone tissue engineered constructs based on starch polycaprolactone scaffolds. *Biomaterials* **30**, 526–534 (2009).

67. Rabkin, E. and Schoen, F.J. Cardiovascular tissue engineering. *Cardiovasc Pathol* **11**, 305–317 (2002).

68. Kania, R.E., Meunier, A., Hamadouche, M., Sedel, L. and Petite, H. Addition of fibrin sealant to ceramic promotes bone repair: long-term study in rabbit femoral defect model. *J Biomed Mater Res* **43**, 38–45 (1998).

69. Grant, I., Warwick, K., Marshall, J., Green, C. and Martin, R. The co-application of sprayed cultured autologous keratinocytes and autologous fibrin sealant in a porcine wound model. *Br J Plast Surg* **55**, 219–227 (2002).

70. Kalia, P., Blunn, G.W., Miller, J., Bhalla, A., Wiseman, M. and Coathup, M.J. Do autologous mesenchymal stem cells augment bone growth and contact to massive bone tumor implants? *Tissue Eng* **12**, 1617–1626 (2006).

71. Kalia, P., Coathup, M.J., Oussedik, S., Konan, S., Dodd, M., Haddad, F.S. and Blunn, G.W. Augmentation of bone growth onto the acetabular cup surface

Vascularisation of tissue-engineered constructs 101

using bone marrow stromal cells in total hip replacement surgery. *Tissue Eng Part A* **15**, 3689–3696 (2009).

72. Marx, G. and Mou, X. Characterizing fibrin glue performance as modulated by heparin, aprotinin, and factor XIII. *J Lab Clin Med* **140**, 152–160 (2002).

73. Neuss, S., Schneider, R.K., Tietze, L., Knuchel, R. and Jahnen-Dechent, W. Secretion of fibrinolytic enzymes facilitates human mesenchymal stem cell invasion into fibrin clots. *Cells Tissues Organs* **191**, 36–46 (2004).

74. Davidson, B.R., Burnett, S., Javed, M.S., Seifalian, A., Moore, D. and Doctor, N. Experimental study of a novel fibrin sealant for achieving haemostasis following partial hepatectomy. *Br J Surg* **87**, 790–795 (2000).

75. Spotnitz, W.D., Dalton, M.S., Baker, J.W. and Nolan, S.P. Reduction of perioperative hemorrhage by anterior mediastinal spray application of fibrin glue during cardiac operations. *Ann Thorac Surg* **44**, 529–531 (1987).

76. Belboul, A., Dernevik, L., Aljassim, O., Skrbic, B., Rådberg, G. and Roberts, D. The effect of autologous fibrin sealant (Vivostat) on morbidity after pulmonary lobectomy: a prospective randomised, blinded study. *Eur J Cardiothorac Surg* **26**, 1187–1191 (2004).

77. Bonucci, E., Marini, E., Valdinucci, F. and Fortunato, G. Osteogenic response to hydroxyapatite-fibrin implants in maxillofacial bone defects. *Eur J Oral Sci* **105**, 557–561 (1997).

78. Hojo, M., Inokuchi, S., Kidokoro, M., Fukuyama, N., Tanaka, E., Tsuji, C., Miyasaka, M., Tanino, R. and Nakazawa, H. Induction of vascular endothelial growth factor by fibrin as a dermal substrate for cultured skin substitute. *Plast Reconstr Surg* **111**, 1638–1645 (2003).

79. Abiraman, S., Varma, H.K., Umashankar, P.R. and John, A. Fibrin glue as an osteoinductive protein in a mouse model. *Biomaterials* **23**, 3023–3031 (2002).

80. Kleinman, H.K. and Martin, G.R. Matrigel: basement membrane matrix with biological activity. *Semin Cancer Biol* **15**, 378–386 (2005).

81. Risau, W. in *Growth Factors and Matrix Influences* 58 (Karger, Basel, 1991).

82. Urbich, C. and Dimmeler, S. Endothelial progenitor cells. *Circ Res* **95**, 343–353 (2004).

83. Hirschi, K.K., Skalak, T.C., Peirce, S.M. and Little, C.D. Vascular assembly in natural and engineered tissues. *Ann NY Acad Sci* **961**, 223–242 (2002).

84. Nomi, M., Atala, A., Coppi, P.D. and Soker, S. Principals of neovascularization for tissue engineering. *Mol Aspects Med* **23**, 463–483 (2002).

85. Clarke, S.A., Hoskins, N.L., Jordan, G.R. and Marsh, D.R. Healing of an ulnar defect using a proprietary TCP bone graft substitute, JAX™, in association with autologous osteogenic cells and growth factors. *Bone* **40**, 939–947 (2007).

86. Geiger, F., Lorenz, H., Xu, W., Szalay, K., Kasten, P., Claes, L., Augat, P. and Richter, W. VEGF producing bone marrow stromal cells (BMSC) enhance vascularization and resorption of a natural coral bone substitute. *Bone* **41**, 516–522 (2007).

87. Tammela, T., Enholm, B., Alitalo, K. and Paavonen, K. The biology of vascular endothelial growth factors. *Cardiovasc Res* **65**, 550–563 (2005).

88. Gerber, H.-P., McMurtrey, A., Kowalski, J., Yan, M., Keyt, B.A., Dixit, V. and Ferrara, N. Vascular endothelial growth factor regulates endothelial cell survival

102   Standardisation in cell and tissue engineering

through the phosphatidylinositol 3'-kinase/Akt signal transduction pathway. *J Biol Chem* **273**, 30336–30343 (1998).

89. Fong, G.-H., Rossant, J., Gertsenstein, M. and Breitman, M.L. Role of the Flt-1 receptor tyrosine kinase in regulating the assembly of vascular endothelium. *Nature* **376**, 66–70 (1995).

90. Neufeld, G., Cohen, T., Gengrinovitch, S. and Poltorak, Z. Vascular endothelial growth factor (VEGF) and its receptors. *FASEB J* **13**, 9–22 (1999).

91. Drake, C.J. and Little, C.D. Exogenous vascular endothelial growth factor induces malformed and hyperfused vessels during embryonic neovascularization. *Proc Nat Acad Sci* **92**, 7657–7661 (1995).

92. Wernike, E., Montjovent, M.O., Liu, Y., Wismeijer, D., Hunziker, E.B., Siebenrock, K.A., Hofstetter, W. and Klenke, F.M. VEGF incorporated into calcium phosphate ceramics promotes vascularisation and bone formation *in vivo*. *Eur Cell Mater* **19**, 30–40 (2010).

93. Deckers, M.M.L., Karperien, M., van der Bent, C., Yamashita, T., Papapoulos, S.E. and Löwik, C.W. Expression of vascular endothelial growth factors and their receptors during osteoblast differentiation. *Endocrinology* **141**, 1667–1674 (2000).

94. Basilico, C. and Moscatelli, D. The Fgf family of growth factors and oncogenes, in *Advances in Cancer Research* (eds George, F.V.W. and George, K.) 115–165 (Academic Press, 1992).

95. Cross, M.J. and Claesson-Welsh, L. FGF and VEGF function in angiogenesis: signalling pathways, biological responses and therapeutic inhibition. *Trend Pharmacol Sci* **22**, 201–207 (2001).

96. Grant, D.S., Kleinman, H.K., Goldberg, I.D., Bhargava, M.M., Nickoloff, B.J., Kinsella, J.L., Polverini, P. and Rosen, E.M. Scatter factor induces blood vessel formation *in vivo*. *Proc Nat Acad Sci* **90**, 1937–1941 (1993).

97. Massagué, J., Blain, S.W. and Lo, R.S. TGFβ signaling in growth control, cancer, and heritable disorders. *Cell* **103**, 295–309 (2000).

98. Dickson, M.C., Martin, J.S., Cousins, F.M., Kulkarni, A.B., Karlsson, S. and Akhurst, R.J. Defective haematopoiesis and vasculogenesis in transforming growth factor-beta 1 knock out mice. *Development* **121**, 1845–1854 (1995).

99. Ferrari, G., Cook, B.D., Terushkin, V., Pintucci, G. and Mignatti, P. Transforming growth factor-beta 1 (TGF-β1) induces angiogenesis through vascular endothelial growth factor (VEGF)-mediated apoptosis. *J Cell Physiol* **219**, 449–458 (2009).

100. Roberts, A.B., Sporn, M.B., Assoian, R.K., Smith, J.M., Roche, N.S., Wakefield, L.M., Heine, U.I., Liotta, L.A., Falanga, V. and Kehrl, J.H. Transforming growth factor type beta: rapid induction of fibrosis and angiogenesis in vivo and stimulation of collagen formation in vitro. *Proc Nat Acad Sci* **83**, 4167–4171 (1986).

101. Tonnesen, M.G., Feng, X. and Clark, R.A.F. Angiogenesis in wound healing. *J Investig Dermatol Symp Proc* **5**, 40–46 (2000).

102. Pollman, M.J., Naumovski, L. and Gibbons, G.H. Vascular cell apoptosis: cell type–specific modulation by transforming growth factor-ß1 in endothelial cells versus smooth muscle cells. *Circulation* **99**, 2019–2026 (1999).

103. Walshe, T.E. TGF-β and microvessel homeostasis. *Microvasc Res* **80**, 166–173 (2010).

104. Pardali, K. and Moustakas, A. Actions of TGF-β as tumor suppressor and pro-metastatic factor in human cancer. *Biochim Biophys Acta (BBA) – Rev Cancer* **1775**, 21–62 (2007).
105. Battegay, E.J., Rupp, J., Iruela-Arispe, L., Sage, E.H. and Pech, M. PDGF-BB modulates endothelial proliferation and angiogenesis *in vitro* via PDGF beta-receptors. *J Cell Biol* **125**, 917–928 (1994).
106. Pipsa Saharinen, K.A. The yin, the yang, and the Angiopoietin-1. *J Clin Invest* **121**, 2151–2159 (2011).
107. Shweiki, D., Itin, A., Soffer, D. and Keshet, E. Vascular endothelial growth factor induced by hypoxia may mediate hypoxia-initiated angiogenesis. *Nature* **359**, 843–845 (1992).
108. Semenza, G.L. Targeting HIF-1 for cancer therapy. *Nat Rev Cancer* **3**, 721–32 (2003).
109. Forsythe, J.A., Jiang, B.H., Iyer, N.V., Agani, F., Leung, S.W., Koos, R.D. and Semenza, G.L. Activation of vascular endothelial growth factor gene transcription by hypoxia-inducible factor 1. *Mol Cell Biol* **16**, 4604–4613 (1996).
110. Zhao, Q., Shen, X., Zhang, W., Zhu, G., Qi, J. and Deng, L. Mice with increased angiogenesis and osteogenesis due to conditional activation of HIF pathway in osteoblasts are protected from ovariectomy induced bone loss. *Bone* **50**, 763–770 (2011).
111. Levenberg, S., Golub, J.S., Amit, M., Itskovitz-Eldor, J. and Langer, R. Endothelial cells derived from human embryonic stem cells. *Proc Nat Acad Sci* **99**, 4391–4396 (2002).
112. Luong, E. and Gerecht, S. Stem cells and constructs for vascularizing tissue engineering constructs. *Adv Biochem Eng Biotechnol* **114**, 129–172 (2009).
113. Zelzer, E., Glotzer, D.J., Hartmann, C., Thomas, D., Fukai, N., Soker, S. and Olsen, B.R. Tissue specific regulation of VEGF expression during bone development requires Cbfa1/Runx2. *Mechanisms of Development* **106**, 97–106 (2001).
114. Chen, W.-J., Jingushi, S., Aoyama, I., Anzai, J., Hirata, G., Tamura, M. and Iwamoto, Y. Effects of FGF-2 on metaphyseal fracture repair in rabbit tibiae. *J Bone Miner Metabol* **22**, 303–309 (2004).
115. Pun, S., Dearden, R.L., Ratkus, A.M., Liang, H. and Wronski, T.J. Decreased bone anabolic effect of basic fibroblast growth factor at fatty marrow sites in ovariectomized rats. *Bone* **28**, 220–226 (2001).
116. Matsuda, T. Recent progress of vascular graft engineering in Japan. *Artif Organs* **28**, 64–71 (2004).

Part II
Standards and protocols in cell and tissue engineering

6

Standards in cell and tissue engineering

P. TOMLINS, Consultant, UK

DOI: 10.1533/9780857098726.2.107

Abstract: This chapter describes current best practice in aspects of tissue engineering ranging from characterization of starting materials to assessing cell behaviour. Technical challenges to developing standards in this area are due in part to the range of materials, cell types, manufacturing and delivery routes that are used to produce products that can be used in many different applications. Guidance documents describing measurements that can be made and techniques to obtain them are a valuable source of information for specialists wishing to gain knowledge beyond their expertise in this multi-disciplinary subject.

Key words: standards, tissue engineering, tissue scaffolds, materials, characterization.

6.1 Introduction

Tissue engineering and regenerative medicine are terms that are often used synonymously to mean replacement or regeneration of damaged or diseased tissue. However, the definitions show that there are some subtle differences between the two terms:

1. *Regenerative medicine:* replaces or regenerates human cells, tissues or organs to restore or establish normal function (Mason and Dunnill, 2008).
2. *Tissue engineering:* use of a combination of cells, engineering, materials and methods to manufacture *ex vivo* living tissues and organs that can be implanted to improve or replace biological functions (BSI PAS 84:2008, (2008)). *Note: Usually through the use of scaffolds for restoration or regeneration of tissues or organs.*

Tissue engineering is, therefore, one aspect of regenerative medicine. The number of papers published and patents awarded in tissue engineering has mushroomed over the past 30 years or so, reflecting the intense global research activity that has happened in this new area of medicine. A brief search of Google patents using the search term 'tissue engineering'

107

© Woodhead Publishing Limited, 2013

108   Standardisation in cell and tissue engineering

provides just over 21 000 hits for applications that range from the development of new types of scaffold, through to cell encapsulation and novel cell delivery systems, for example, self-assembling hydrogels. Despite this high level of activity, translation of this knowledge into commercially available products has been much slower (Johnson *et al.*, 2011). Factors influencing the rate at which tissue-engineered medical products (TEMPs) are being commercialized include the unrelenting challenges of securing finance in challenging economic times, and selling the new technologies into a relatively conservative market that is becoming increasingly price-sensitive. The delay has also been attributed in part to the fact that many TEMPs fall outside the existing regulatory frameworks and standards that cover medical devices and medicines (Plagnol *et al.*, 2009; Messenger and Tomlins, 2011).

The development of globally accepted standards that describe test protocols can help in the process of product commercialization. These documents cover characterization of the materials used in tissue engineering, scaffold structures and the performance of ensembles that contain living cells. Tests that have been carried out according to a consensus-approved methodology which extends beyond individuals, research groups and organizational procedures tend to produce results that are much less operator/equipment-dependent. Having protocols in place to ensure that tests can be carried out reproducibly, that is, not only by the same operator using the same equipment but also different operators in different laboratories using a range of experimental set-ups, is key to ensuring consistency of data. Consistent data provide a measure of quality assurance for suppliers as well as regulatory bodies and potential customers. Performing tests to an agreed standard can also be of value to manufacturers and suppliers if the test data are ever queried, and enhances the comparability of research findings. The development of standards that describe well-defined test protocols lags behind the development of new technologies. This is primarily because it can be difficult to identify the most appropriate test method to determine some characteristic of a material or structure in a rapidly developing field, as will be discussed later in this chapter. It can also take time to identify the most useful things to measure. Compliance with standards, however, can be used as a mechanism to ensure compliance with legal requirements, such as those of the European Commission Medical Directive 93/42/EEC, 1993.

Compliance with technical standards forms only part of the picture. The ISO 9000 family of quality standards (ISO 9000, 2005) provides a much larger framework for dealing with the mechanics of complying with procedures, storing information etc. and is widely used throughout industry. ISO 13485 (2003) is a similar quality standard to the ISO 9000 portfolio, but it specifically addresses the management system requirements for manufacturing

© Woodhead Publishing Limited, 2013

Table 6.1 The national and international standards bodies active in drafting documents that support tissue engineering and regenerative medicine

Organization	Primary coverage	Key committees	Website URL
BSI	National	RGM/1	www.bsi.org
ASTM International	Global	F04.41 to F04.46	www.astm.org
ISO	Global	TC 150/SC7	www.iso.org
CEN	European	TC316	www.cen.eu

medical devices. This includes inspection requirements, sterilization, traceability, risk analysis and controls for ensuring product safety.

6.2   How and by whom are standards produced?

Standards are published by organizations that operate at levels ranging from national, for example, the British Standards Institute (BSI) through regional, for example, the Committee for European Standardization (CEN),[1] to global, for example, the International Organisation for Standardisation (ISO) (see Table 6.1). The publications produced by these organizations range from guidance documents that have an educational focus to procedural documents that provide detailed guidance on how to carry out specific tests, including sample preparation and reporting of results. Drafting of the standards typically requires one or more authors to produce a document that is then extensively peer reviewed by a panel of national or international experts. Guidance documents tend to have a shorter timescale to publication compared with procedural texts which require a body of statistically robust test data to be produced prior to being approved, a process that can take years. The technical committees (TCs) listed in Table 6.1 provide further up-to-date information on standards that are being developed.

The division of standards bodies into national, regional and global is to some extent artificial as there are many examples where national standards have developed into international documents; for example, the ISO 9000 family of quality standards originates from British standard, BS5750 (BS 5750:pt 1, 1979).

6.3   The importance of an agreed lexicon

A fundamental step in preparing a portfolio of standards for tissue engineering is to establish a common lexicon of terms and definitions that can be

---

[1] Note: Standards developed by CEN automatically become national standards in 31 countries.

110   Standardisation in cell and tissue engineering

used by all the standards bodies. Establishing consistent definitions provides clarity that helps to minimize any potential misunderstandings that might arise within the community, especially in a multi-disciplinary area such as tissue engineering. A good example of this is the characterization of tissue scaffolds, where the terms macropores, mesopores and micropores refer to quite different length scales to those used by chemists to describe the structure of catalysts. Standards that cover the terminology of tissue engineering include:

- ASTM F2211-04: Standard specification for general classification for tissue-engineered medical products.
- BSI PAS 84-2008: Regenerative medicine – glossary.

Occasionally minor amendments to the definitions of terms that take into account new findings or provide a more succinct description can make the task of harmonization across a standards portfolio more difficult. A route around this issue that is used by ASTM International is to place all definitions in an online database for all of its standards (ASTM Dictionary of Engineering Science and Technology).

6.4   Drivers for standardization

Standards are primarily driven by regulatory and policy needs and can provide a mechanism for ensuring product compliance with legal requirements. In Europe, tissue-engineered medical products that are cell-free need to comply with the Medical Devices Directive (European Commission Medical Directive 93/42/EEC, 1993). This is a wide-ranging directive that covers products from orthopaedic implants through to spectacles and beyond. The UK Medicines and Healthcare products Regulatory Agency (MHRA, http://www.mhra.gov.uk/Howweregulate/Devices/Detailsofharmonisedstandardsunderthemedicaldevicesdirectives/index.htm) provides an excellent overview of the standards that are used to ensure compliance with this directive (MHRA). Products that contain cells are regarded in Europe as medicines and will need to comply with the Advanced Therapy Medicinal Products Directive (1983), which doesn't use standards as a primary means of ensuring compliance. Further information can be found in the Medicines for Human Use Regulations (2010). Aside from the regulatory requirements, ASTM International is developing standards for characterizing cells and their behaviour, and these will be discussed in more detail below. Of course, there is no reason why standards can't be complied with on a voluntary basis, and standards have been published that are based on good practices developed by companies or academic researchers.

© Woodhead Publishing Limited, 2013

Methods and protocols 111

6.5 How will standards help me?

There are many instances where labs have developed an 'in-house' test method to characterize materials, or to assess cell behaviour or the efficacy of a new treatment and, whilst these are extremely useful, they suffer the disadvantage that no-one, particularly those outside the company, actually knows how 'good' the data are that they produce. On a day-to-day, batch-to-batch basis this isn't really an issue, providing the test method is sensitive enough to pick up any variability that could have a negative impact on product performance, and the results are independent of the operator. The sensitivity of the method to these variables should be understood and minimized to ensure that the data produced is of the highest quality. It has also been suggested that the introduction of standards can impede innovation and add an unnecessary cost burden to start-up companies. However, sound arguments can also be made that show that compliance with standards can lead to cost savings through production of consistent, well-characterized materials or structures.

6.6 What standards currently exist in tissue engineering?

We have already seen that some tissue-engineered medical products (TEMPs) are classed as medical devices and are covered by the same standards that are used for, for example, the manufacture and testing of orthopaedic implants. The ISO10993 series of standards, comprising 20 parts (ISO 10993), addresses best practice in the biological evaluation and testing of medical devices including, amongst others, tests for toxicity (parts 3, 11 and 20), interactions with blood (part 4) *in vitro* cytotoxicity (part 5), tests for local effects after implantation (part 6), and evaluation of degradation products (parts 9,13,14 and 15). In addition, ASTM International has published F748 'Standard Practice for selecting biological test methods for materials and devices' (2010) that recommends generic biological test methods for materials and devices according to end-user applications. Potential safety risks associated with bacteria, fungi, mycoplasma, viruses, endotoxins, transmissible spongiform encephalopathies (TSEs) and other pyrogens, parasitic organisms and their by-products are also addressed in ASTM F2383 (2011). The guide does, however, exclude products containing live cells or tissues. However, there are aspects of tissue engineering that require the development of new standards or adaptation of documents that cover similar materials or products that are used in other applications. The key areas that need to be addressed in developing a TEMP are:

- Characterization of the starting materials used, which also includes cells.
- Characterization of tissue scaffolds to ensure consistent manufacture and performance.

© Woodhead Publishing Limited, 2013

112 Standardisation in cell and tissue engineering

- Assessment of the performance of products that may consist of both a scaffold and cells.
- Storage and handling of TEMPs.

Standards that can support best practice and quality in any of these areas should be developed in such a way that their introduction does not have any perceived negative impact on innovation. There are also many instances, as discussed below, where it is by no means clear as to which tests are the most appropriate to perform on a material/structure or product. One way round these limitations is to produce guidance documents that discuss the relative merits of different techniques/approaches to testing without being prescriptive. Examples of this type of document are called Standard Guides by ASTM International, and can form the basis of future more prescriptive standards. The guides are also useful for experts in one aspect of tissue engineering wishing to extend their knowledge to areas that they are less familiar with.

6.7 Characterization of biomaterials and biomolecules

There are many examples of tissue scaffolds that are made from naturally derived materials harvested from both plants, for example, alginates (Wang *et al.*, 2012), and animals, for example, collagen, fibroin (Glowacki and Mizuno, 2008; Liu *et al.*, 2011) and combinations of the two, for example, chitosan-alginate (Li *et al.*, 2005), as well as natural-synthetic combinations, for example, alginate-bioglass (Mouriño *et al.*, 2010). Materials that are derived from natural sources tend to be much more variable in terms of their composition, especially when compared with synthetic materials, for example, in the alginates the ratio of poly(mannuronic) acid to poly(guluronic) acid in the polymer depends on the location of the algae, species and growing season (Haug *et al.*, 1969; South, 1979; Miller, 1996). A number of standard guides have been published which specifically cover the characterization of the starting materials used to produce scaffolds, particularly those of natural origin (F2027-08, 2008; F2212-09, 2009). ASTM F2027, in particular, provides an overview of characterization and testing of raw materials which may be a useful source of information in the absence of a specific material guide.

6.8 Characterization of tissue scaffolds

Guidance has also been published on the characterization of tissue scaffolds: ASTM F2150-07 (2007) provides an extensive list of standards that can be used to evaluate scaffolds. ASTM F2450-10 (2010) gives a more detailed description of scaffold characterization and assesses the different techniques

© Woodhead Publishing Limited, 2013

Methods and protocols 113

Table 6.2 Methods that can be used to characterize tissue scaffolds and the information that they can provide

Test modality	Generic technique	Information obtained
Imaging	Micro-computer X-ray tomography Microscopy Magnetic resonance imaging	Pore shape and size, size distribution and porosity
Permeability	Mercury porosimetry	Porosity, total pore surface area, pore diameter, pore size distribution
	Gas flow porometry	Median pore diameter (assuming cylindrical geometry), through pore size and distribution
	Diffusion of markers Darcy permeability coefficient	Permeability
Other	Nuclear magnetic resonance	Pore size and distribution
	Density measurements	Porosity, pore volume

Source: Adapted from ASTM F2450-10 (2010).

that are used, both in terms of the information that they provide and how they can be applied. Typically the methods used to characterize scaffolds can be broken down into those that produce 2D or 3D images of the structure and those that produce some measure based on permeability, Table 6.2.

A fundamental 'problem' with tissue scaffolds is that they are manufactured from materials that range from highly hydrated hydrogels to ceramic matrices, which makes it difficult to identify generic test methods. Furthermore, scaffolds can, depending on how they are manufactured, have complex structures that consist of a combination of different pore types that may also be extensively interlinked. Closed pores (Fig. 6.1) look like bubbles within the structure that are difficult to identify using the methods listed in Table 6.2, although their presence can be inferred from density measurements. However, it can be difficult to differentiate between them and trapped air bubbles in poorly wetted-out scaffolds if the density is measured using a wetting balance. Identifying closed pores is important as they can have a significant effect on the performance of tissue scaffolds by effectively reducing the diffusion distance within the structure for dissolved gases such as oxygen and carbon dioxide.

Blind-end pores (Fig. 6.1) perform a similar function to closed pores and are also not that easy to identify. The through-pores play the most important role in a scaffold, providing conduits for nutrient/waste product/dissolved gas transport and cell colonization. Through-pores can be simple

© Woodhead Publishing Limited, 2013

114 Standardisation in cell and tissue engineering

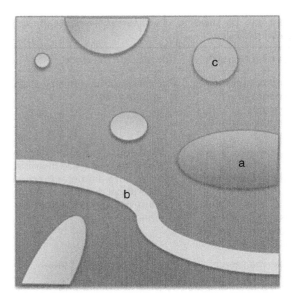

6.1 Tissue scaffolds typically consist of a combination of (a) blind-end, (b) through- and (c) closed pores.

conduits or they may be highly interconnected and may also have apertures in their walls that link to adjacent pores. The dimensions of such ill-defined structures are difficult to determine, although techniques such as mercury porosimetry can be used to obtain a representation of the structure as a distribution of parallel-sided, non-interconnected channels. Perhaps a more useful measure of the performance of this type of complicated structure is to determine its Darcy permeability coefficient (Darcy, 1856; Wang *et al.*, 2010). The Darcy permeability coefficient is widely used to characterize rocks and soils (D4525-08, 2008) and is a measure of the resistance of a porous structure to the flow of fluid through it. It, like most other metrics used in scaffold characterization, has its limitations, that is, the value will be influenced by defects such as cracks in the structure or surface skins that limit fluid penetration, but these 'features' will also be encountered when the scaffolds are used.

Many papers published in tissue engineering use some form of image analysis to provide a measure of scaffold porosity and pore dimensions. This information is often derived from 2D images, such as scanning electron micrographs, but it can be based on 3D micro-computer X-ray tomography images. The key to the successful analysis of images is to define a threshold intensity which serves to differentiate between those parts of an image which consist of material and the space that forms the pores. This is difficult to do using automated methods with 2D scanning electron micrographs, due

Methods and protocols   115

to the extended depth of field that exists, but is relatively straightforward for 3D images, or 2D images that have very limited depth of field. Many methods are available to define the threshold in a more objective way (see, for example, Otsu, 1979), but often the subjective 'by eye' approach works acceptably well. However, using the wrong value for the threshold can significantly change the calculated porosity of a scaffold (Mather *et al.*, 2008). ASTM F2603-06 (2006) provides guidance on how to interpret images of polymer-based scaffolds, discussing different approaches to thresholding and methods for capturing and storing data.

Porosity is widely used as a measure of the volume of free space that exists within a scaffold, but in practice it is not that useful a metric in the absence of any complementary data, that is, percentages of different pore types, pore dimensions, etc. Figure 6.2 shows examples of three structures that have the same degree of porosity. Structure (a) in Fig. 6.2 would not be a particularly useful scaffold since it will be impervious to cells. Structures (b) and (c) could be used as scaffolds but the microenvironments that surround the cells will be quite different, which may affect cell behaviour over an extended timescale.

The surface texture of scaffolds, or indeed implantable biomaterials in general, is a significant factor influencing cell behaviour. Typically surface roughness is reported as a single value, Ra. Ra is an averaging parameter which provides an indication of how 'rough' a 2D profile of the surface is, but no more. Ra is a measure of the mean deviation of the profile peaks and valleys from a mean (ISO 4287, 1997). Although it is widely recognized that this approach is relatively insensitive to different surface textures, it is still extensively used to characterize biomaterials as well as in engineering. In practice, the insensitivity of this metric means that it may not be able to distinguish between surfaces that look quite different from a cell's perspective, which may help to explain some of the variability in results obtained from experiments that involve cells. A more pragmatic, but challenging approach to quantifying surface texture would be to consider the surface from a cell's perspective, that is, what features of the surface would be 'seen' by an object that is usually much less than 100 μm in size, when considering which metrics to use. There are many other parameters that can be used to quantify surface texture, both in 2D and 3D, with some quantifying areal textures. Further information and guidance on how to determine and use these can be found in ASTM F2791-09 (2009).

Despite the important influence of surface texture on cell behaviour, it is not easy to measure surface texture profiles for porous tissue scaffolds due to their small size, often highly convex or concave surfaces and inaccessibility. Scanning electron microscopy (SEM) can be used to examine the surface texture of the struts that form the structure of the scaffold, but these are difficult to quantify, unless 3D images are available. A practical

© Woodhead Publishing Limited, 2013

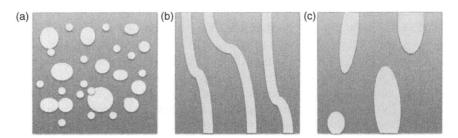

6.2 Scaffolds that have the same porosity can have quite different structures: (a) is composed of closed pores; (b) of through-pores; and (c) of blind-end pores.

workaround this problem is to produce a range of surface textures on planar specimens. These can be manufactured with differing levels of periodicity, randomness, etc. using commercially available textures as templates. These model systems can be used in comparative studies to find out just how important surface texture is in terms of affecting cell behaviour for a given cell/material combination (Deligiannia *et al.*, 2001; Davis *et al.*, 2010). Measuring how well cells attach to surfaces is a key metric for optimizing the chemistry and topography of a given surface, and many approaches have been developed which are discussed in ASTM F2664-07 (2007). From unpublished pre-standardization activity it seems that measurements of the force required to detach cells that have adhered to a surface are susceptible to many factors, including the type of measurement being made; the design of the apparatus; the cell culture history; and operator handling. However, progress is being made towards developing a consensus approach to adhesion measurements that are based on centrifugation (Kaplan *et al.*, 2012).

Cells interact with the adsorbed layer of proteins that rapidly forms when tissue scaffolds are immersed in culture medium containing cells. The concentration and type of protein in the adsorbed layer will no doubt be influenced by nanoscale surface texture, but more importantly, by the surface chemistry and charge. These two aspects of scaffolds have received little attention in the literature and are not currently covered by standards. There are practical challenges in establishing what molecular species are present at the surface and what the charge is on it. Techniques that are surface-sensitive to a depth of a few nanometres, such as X-ray photoelectron spectroscopy (XPS) require a relatively large sample footprint, around 100 μm, which is larger than that of most tissue scaffold strut diameters. Energy dispersive X-ray spectroscopy is often used to identify molecular species present in SEM samples, but the high penetration depth of this technique (around 2 μm) results in it being more a measure of the near-surface composition (Torrisi *et al.*, 2008). Some indication of the surface chemistry can be obtained by using the same cleaning/handling techniques and packaging/storage methods on planar samples, if they can be manufactured, as those used on scaffolds.

Many of the structural characterization techniques require 'dry' scaffolds, for example, gas flow porometry, mercury porosimetry and conventional SEM. However, there are many examples of scaffolds that are used in a highly hydrated state, for example, collagen and alginate. Procedures used for dehydrating these materials are based on a series of ethanol/water mixtures (e.g., Jacchetti *et al.*, 2008) or lyophilization (e.g., Trieu and Qutubuddin, 1994), but there is always some uncertainty as to how well the structure observed relates to that which exists in the hydrated state, due to the susceptibility to structural damage or dimensional changes that can occur during dehydration, such as shrinkage or fracturing. There are many other factors to be considered that are often overlooked when interpreting characterization data obtained from complementary techniques. Examples of these factors include the location of the sample volume that has been characterized in the structure. Some consideration needs to be given to the depth in the scaffold sample at which representative samples are harvested for SEM examination, how many in-plane samples should be taken and from which positions: near an edge, towards the centre or some combination of the two? Further thought needs to be given to how any variability in the structures will be addressed; averaging is an obvious route, but this may effectively mask a degree of anisotropy that has an impact on cell behaviour. Problems due to structural through-thickness anisotropy can be difficult to experimentally distinguish from the adverse effects of diffusion-limited processes such as the build-up of waste metabolic products in cell-seeded scaffolds.

Residual stress in scaffold structures can lead to internal damage or dimensional changes when samples are removed for characterization by, for example, microscopy. Residual stress is generated as a result of constraints being imposed on the sample when it is manufactured. The residual stress builds up in the structure when, for example, a skin is formed that prevents any dimensional changes from occurring naturally, for example, shrinkage in the structure that forms beneath it. This typically occurs in ionic cross-linking of polymers during the production of hydrogels or those processes that form cavities as a result of gas expansion in the structure, for example, supercritical carbon dioxide. Residual stress can also be created as a result of water swelling when a scaffold is immersed in culture medium. A simple test to see if high levels of residual stress are present is to cut a sample open and then look for changes in the sample dimensions.

Recently NIST have produced commercially available scaffolds with a relatively simple lattice structure that can be used as reference materials to assess cell–cell batch variability (RM 8396) and in comparative studies to evaluate measurement protocols.

118   Standardisation in cell and tissue engineering

The degradation behaviour of scaffolds is a key aspect of performance for many applications. Like most other aspects of tissue engineering, degradation behaviour and its impact on cell behaviour are far from simple to understand. In some cases, for example, scaffolds produced from poly(lactides), the degradation mechanism is through hydrolysis in an autocatalytic reaction, that is, the core region degrades at a faster rate than the outer surfaces (Weir *et al.*, 2004; Wu and Ding, 2004). There is some guidance in the standards (ISO 10993: Part 13, 2010) for assessing the degradation behaviour of scaffolds. The protocol is to carry out measurements in a liquid environment at an elevated temperature, effectively shortening the test period, which would otherwise extend to many years at body temperature. Samples of the fluid bath are periodically taken to measure the concentration of degradation products. Published papers that report such measurements demonstrate the challenges faced in trying to standardize this type of measurement; variables include the sample geometry, flask geometry, volume of the fluid bath in which the sample is immersed, frequency at which the fluid bath is changed, sample volume, temperature, number of repeat measurements, sampling interval and degree of mechanical agitation (Agrawal *et al.*, 2000; Yoon and Park, 2001; Pêgo *et al.*, 2003; Hedberg *et al.*, 2005). Some allowance may need to be made for complex porous 3D matrices due to the fact that the measured concentration of degradation products in the fluid that contains the sample may be lower than that present in the core of the structure due to diffusion constraints. Further consideration needs to be given to samples that are enzymatically degraded, for example, collagen, where the enzyme concentration is obviously important. Steps may also need to be taken to limit microbial contamination of the materials being tested. The introduction of an antibiotic as an additional experimental variable, can result in experiments running for many months.

Relating *in vitro* measurements to *in vivo* performance is a challenge, particularly as the volume of free space in a cell-seeded scaffold progressively declines with time. The scaffold may also be subjected to mechanical forces *in vivo*, depending on the application, that would not be considered in an *in vitro* experiment due to movement of the host animal which can change the dynamics of fluid flow into and out of the scaffold. A direct consequence of fluid flow is to change the dynamics of diffusion gradients for metabolites, waste products and potentially pH. There are potential difficulties in relating the degradation behaviour of bulk samples to porous scaffolds due to differences in the surface area to volume ratio and the creation of microenvironments. The surface area to volume ratio will be very important for surface degradation and will affect diffusion-limited processes such as hydrolysis.

Hydrogels are extensively used as tissue scaffolds as well as cell delivery vehicles. Environmentally sensitive polymer solutions that form hydrogels

© Woodhead Publishing Limited, 2013

Methods and protocols 119

Table 6.3 Key factors and aspects that need to be considered when characterizing hydrogels

Key factors for hydrogel characterization				
Aspect	Biological properties	Kinetics	Physical and chemical stability	Mass transport
	Biocompatibility	Gelling time	Environmental stability	Cell migration
	Adventitious agents	Swelling rate	Mechanical properties	Transport of nutrients and waste
		Matrix degradation	Cell encapsulation	Release rate of bioactive agents

Source: From reference F2900-11 (2011).

as a result of a change in pH or temperature, for example, form ideal systems for delivering cells into the body and encapsulating them *in situ*. ASTM F2900-11 (2011) is a standard guide that describes test methods for characterizing hydrogels. Table 6.3 gives an overview of the topics discussed in this document, which range from the kinetics of hydrogel formation to determining the release rates of bioactive agents.

Developing best practice guidance to track the time-dependent release of biomolecules, such as growth factors, from scaffolds is subject to many of the same factors that affect degradation testing, for example, sample geometry, volume of supporting fluid and mechanical agitation. ASTM International is currently considering standard guides for both degradation and *in vitro* release of bioactive agents from scaffolds.

6.9 Characterization of cell-seeded scaffolds

Selecting a suitable scaffold structure manufactured from a material that has suitable mechanical properties and geometry for a given cell type that will continue to perform its function over an extended period of time is a significant challenge. Such requirements need to be met in an environment where not only will the cell density increase with time, but the scaffold itself may also be continuously degrading. Failure of the scaffold system to provide a suitable environment can result in death of the cells in the core of the structure, loss of phenotype and/or senescence. The method used to initially seed the scaffold can also have a bearing on its subsequent performance, that is, whether the cells are passively delivered to the structure through pipetting or actively 'pushed' into it using mild centrifugation. ASTM F2739-08 (2008) can help in optimizing the seeding method and in understanding cell

© Woodhead Publishing Limited, 2013

120   Standardisation in cell and tissue engineering

behaviour in a 3D matrix, since it describes test methods that can differentiate between viable and non-viable cells and whether such cells are proliferating or not. These test methods, carried out *in vitro*, can be indicative of *in vivo* performance, although the latter brings in more variables that may be difficult to control or understand, such as uncontrolled mechanical stimulation. Cell viability can also be assessed in cell-seeded scaffolds that have been cultured in an *in vivo* environment to confirm the indications obtained from *in vitro* measurements. Whilst cell viability is a measure of the 'health' of cells, the test methods described in ASTM F2739-08 are not suitable for assessing whether the viable cells are undergoing apoptosis or are under stress. The guide recommends additional measurements of, for example, protein expression and cell morphology to complement the viability data.

ASTM standard guides are available for assessing implantable devices that are intended to repair or regenerate articular cartilage (F2451-10, 2010), and the pre-clinical *in vivo* evaluation of critical-size segmental bone defects (F2721-09, 2009).

6.10   Manufacture, processing and storage

Consistency in TEMPs can only be achieved through the development of robust protocols for manufacturing scaffolds, both harvesting and handling of cells, and storing the resultant product. ASTM F2210-02 (2010) describes methods for processing cells, tissues and organs for use in TEMPs, and F2386-04 outlines procedures for preservation of TEMPs to maximize the retention of desirable characteristics. ASTM F2315-11 (2011) describes test methods for immobilizing or encapsulating living cells in beads of alginate gels and provides a list of the parameters that need to be controlled. Many of the standards developed in the medical device field may also be of assistance to manufacturers, such as the ISO 13408 package of standards (2003–2008) that describe aseptic processing of health care products.

Sterilization is an important issue, particularly for hydrogels or scaffolds that contain bioactive compounds. Using gamma irradiation is likely to cause damage in polymer-based scaffolds and to bioactive compounds such as growth factors. Ethylene oxide can also be problematic to use as a sterilizing agent in tissue engineering due to carcinogenic residues. It may be that the hydrogel has to be manufactured under aseptic conditions or contain an antimicrobial agent, as discussed in ASTM F2900-11 (2011).

Cells and cell-containing scaffolds can be stored using a variety of methods including hypothermic preservation, vitrification and freezing. Each of these methods consists of a series of steps that can have an adverse effect on the behaviour of the cells after recovery and/or the scaffold characteristics, for example, the rates of cooling and warming or freezing/thawing and use of cryoprotectants. Obvious examples include changes to the

© Woodhead Publishing Limited, 2013

Methods and protocols 121

structure of hydrogels due to ice damage during freezing or loss of water during dehydration. ASTM F2386-04 (2004) outlines procedures that can be used to minimize the negative impact that these factors can have on TEMPs and refers to other guidance documents for environmental tracking during transport. Further guidance can be found in documents that cover cell banking, for example, those produced by the US Food and Drug Administration (FDA, 2010).

6.11 Characterization of cells and cell–surface interactions

ASTM International is in the process of developing standards that are focused on identifying metrics for characterizing cell behaviour for single cells as well as on a cell-to-cell basis and for cell–surface interactions. There is tremendous potential in these standards, especially if robust metrics can be derived that could enable quantitative assessments of the cell lines that have been grown in different laboratories for many years, for example, mouse mg63 cells. These cells and many other lines have been the subject of many publications, but given the sensitivity of cells to their culture conditions, comparisons between cells cultured in different laboratories are questionable, as indeed are comparisons over time within a single laboratory due to differences in operator and culture conditions, such as changes in the composition of the culture medium.

Before attempting to characterize cell behaviour, it will be necessary to establish guidelines for cell culture conditions using well-defined media and ideally, a reference cell-line. Establishing a reference cell-line requires the development of an agreed protocol for selecting, storing and characterization of cells, some of which exists within, for example, the FDA guidelines on cell therapy (FDA, 2010). Additional consideration is being given to development of a standard protocol for preparing, storing and handling extra-cellular matrix materials, for example, fibrillar collagen films to elicit an expected response from reference cells cultured under well-defined conditions in a well-defined medium. Achieving this goal is almost a pre-requisite for the development of further best practice guides that will cover cell–cell signalling, cell–2D surface interactions and cell behaviour in a 3D matrix. A lot of pre-standardization activity will be required to determine the appropriate level of detail for adequate characterization of the materials, surfaces and structures to create a sound foundation for more robust studies of cell behaviour.

Some of the metrics used to characterize cell–surface interactions are concerned with developing a best practice approach to determining cell spread area. Guidance on the photostability and potential cytotoxicity of the fluorophores used to image cells and sub-cellular components is also

© Woodhead Publishing Limited, 2013

122 Standardisation in cell and tissue engineering

being developed. Contrast agents are extensively used to image whole cells as well as cellular components, for example, the cell membrane, and nucleus in both 2D and 3D studies. A key consideration with these materials is understanding their limitations, that is, are they cytotoxic to the cell, and if so, under what conditions? Some fluorophores, for example, produce reactive oxygen species when excited by intense laser light that are toxic to cells (phototoxicity), and become less fluorescently intense (photobleached) with time. These effects can be easily encountered in laser scanning confocal imaging of cell-seeded scaffolds; consequently those cells that are nearest the surface that have been photobleached due to the intensity and duration of the laser exposure appear to be smaller than those located deeper within the structure. Reducing the intensity of the laser in a confocal microscope may not improve matters since this action will also reduce the signal-to-noise ratio of the image. A standard guide that describes the different types of fluorophores available and their pros and cons is currently being developed within ASTM International. Fluorescence is also extensively used in cell-sorting machines to segregate cells, for example, stem cells. The purity of the harvest depends, at least in part, on the drop size used, which in turn is influenced by the sensitivity of the fluorescence detection system and the intensity of the fluorophore tags.

The cell signalling group that has been recently formed within ASTM International aims to develop good practice guidance documents that will address such topics as gap junction intracellular communications and early indicators of stress, that is, mutagenicity, inflammation and irritation. Such guidance will provide much greater understanding of cell–material and cell–cell interactions that will no doubt lead to the development of TEMPs and implantable medical devices superior to those available today.

6.12 Conclusion

The objective of developing standard protocols or reference materials for tissue engineering applications is to ensure the consistency of products which is fundamental for market acceptance, patient safety and profitability. Tissue-engineered medical products in their most complicated form consist of a cell-seeded matrix. Over time the number of cells will increase and the scaffold will degrade or be re-modelled. The materials used to form the matrix vary significantly, that is, from ceramics through to hydrogels, and may be sourced from natural materials, such as seaweeds, or chemically synthesized. Standards for tissue-engineered products could therefore potentially cover:

- Manufacture of the raw material, for example, extraction of alginates from seaweed and sterilization (if applicable), storage and development of a 'use-by date'.

© Woodhead Publishing Limited, 2013

Methods and protocols 123

- Manufacture of scaffolds, for example, sintering of ceramics, solvent leaching from polymer-based matrices, use of rapid prototyping technology.
- Characterization of scaffolds (e.g. permeabilty, pore size distribution).
- Degradation behaviour of the scaffold, potentially both *in vitro* and *in vivo* mechanical performance (including time-dependent changes), all of which accommodate differences in sample geometry.
- Mechanical properties, for example, Young's modulus in tension and in compression, strain distribution within the matrix and how these may change with time in a degrading structure.

A similar suite of standards could also be potentially prepared for cells that would include harvesting protocols, storage and handling and methods for objectively assessing cell behaviour. Clearly an enormous amount of work could be done, if technically feasible, to standardize all elements of tissue-engineered products, but most of this is probably not necessary. At this stage of the development of this new technology it is probably sufficient to produce standard guides highlighting the measurements that could be made as well as the pros and cons of the techniques that are used to make them, until the key factors that impact on product quality and safety have been identified.

6.13 References

Advanced Therapy Medicinal Products Directive 2001/83/EC (1983). Since updated, see Regulation (EC) No 1394/2007.
Agrawal CM, McKinney JS, Lanctot D and Athanasiou KA (2000). Effects of fluid flow on the in vitro degradation kinetics of biodegradable scaffolds for tissue engineering. *Biomaterials*, **21**(23): 2443–2452.
ASTM Dictionary of Engineering Science and Technology. www.astm.org.
ASTM (2011). F2900-11 Standard Guide for Characterization of Hydrogels used in Regenerative Medicine. www.astm.org.
BSI (1979). BS5750 part 1: Quality systems. Specification for design, manufacture and installation. www.bsi.org.
BSI (2008). BSI PAS 84:2008 Regenerative Medicine – Glossary, www.bsi.org.
Darcy, H (1856). *Les fontaines publiques de la ville de Dijon*, Dalmont, Paris.
Davies J, Lam JKW, Tomlins PE and Marshall D (2010). An in vitro multi-parametric approach to measuring the effect of implant surface characteristics on cell behaviour. *Biomedical Materials*, **5**: 15002–15008 (DOI: 10.1088/1748-6041/5/1/015002).
D4525-08 (2008). Standard Test Method for Permeabilty of Rocks by Flowing Air. www.astm.org.
Deligiannia DD, Katsalaa, N Ladasb, S Sotiropouloub D, Amedeec, J and Missirlisa, YF (2001). Effect of surface roughness of the titanium alloy Ti–6Al–4V on human bone marrow cell response and on protein adsorption. *Biomaterials*, **22**(11): 1241–1251.

© Woodhead Publishing Limited, 2013

124 Standardisation in cell and tissue engineering

European Commission (2003). *Medical Device Directive* (Council Directive 93/42/EEC of 14 June 1993 concerning medical devices, OJ No L 169/1 of 1993-07-12).

F748-06 (2010). Standard Practice for Selecting Generic Biological Test Methods for Materials and Devices. www.astm.org.

F2027-08 (2008). Standard Guide for Characterization and Testing of Raw or Starting Biomaterials for Tissue Engineered medical Products. www.astm.org.

F2150-07 (2007). Standard Guide for Characterization and Testing of Biomaterial Scaffolds Used in Tissue Engineered Medical Products. www.astm.org.

F2210-2 (2010). ASTM F2210 – 02(2010) Standard Guide for Processing Cells, Tissues, and Organs for Use in Tissue Engineered Medical Product. www.astm.org.

F2211-04. (2004). Standard Classification for Tissue Engineered Medical Products (TEMPs). www.astm.org.

F2212-09 (2009). Standard Guide for Characterization of Type I Collagen as Starting Material for Surgical Implants and Substrates for Tissue Engineered Medical Products (TEMPs). www.astm.org.

F2315-11 (2011). Standard Guide for the Immobilization or Encapsulation of Living Cells or Tissue in alginate Gels. www.astm.org.

F2383-11 (2011). Standard Guide for Assessment of Adventitious Agents in Tissue Engineered Medical Products (TEMPs). www.astm.org.

F2386-04 (2004). Standard Guide for Preservation of Tissue Engineered Medical Products. www.astm.org.

F2450-10 (2010). Standard Guide for Assessing Microstructure of Polymeric Scaffolds for Use in Tissue Engineered Medical Products. www.astm.org.

F2451-10 (2010). Standard Guide for in vivo Assessment of Implantable Devices Intended to Repair or Regenerate Articular Cartilage. www.astm.org.

F2603-06 (2006). Standard Guide for Interpreting Images of Polymeric Tissue Scaffolds. www.astm.org.

F2664-07 (2007). Standard Guide for assessing the attachment of cells to biomaterial surfaces by Physical Methods.

F2721-09 (2009). Standard Guide for Pre-clinical in vivo Evaluation in Critical Size Segmental Bone Defects. www.astm.org.

F2739-08 (2008). Standard Guide for Quantitating Cell Viability within biomaterial scaffolds. www.astm.org.

F2791-09 (2009). Standard Guide for Assessment of Surface Texture of Non-Porous Biomaterials in Two Dimensions. www.astm.org.

FDA (2010). Characterization and Qualification of Cell Substrates and Other Biological Materials Used in the Production of Viral Vaccines for Infectious Disease Indications. www.fda.gov/downloads/biologicsbloodvaccines/.../ucm202439.pdf.

Glowacki J and Mizuno S. (2008). Collagen scaffolds for tissue engineering. *Biopolymers*, **89**(5): 338–344.

Haug A, Larsen B and Baardseth E (1969). Comparison of the constitution of alginates from different sources. *Proc VI Int Seaweed Symp*, 443–451.

Hedberg EL, Kroese-Deutman HC, Shih CK, Crowther RS, Carney DH, Mikos AG and Jansen JA (2005). *In vivo* degradation of porous poly(propylene fumarate)/poly(DL-lactic-co-glycolic acid) composite scaffolds. *Biomaterials*, **26**(22): 4616–4623. Epub 18 January 2005.

© Woodhead Publishing Limited, 2013

ISO (1997). ISO 4287:1997 Geometrical Product Specifications (GPS) – Surface texture: Profile method – Terms, definitions and surface texture parameters. www.iso.org.

ISO (2005). ISO 9000:2005. Quality management systems – fundamentals and vocabulary.ISO (2008) ISO 9001:2008. Quality management systems – requirements. www.iso.org.

ISO 10993 parts 1-20. Details of these documents can be found at www.iso.org.

ISO (2010). ISO 10993-13: 2010. Biological evaluation of medical devices – Part 13: Identification and quantification of degradation products from polymeric medical devices. www.iso.org.

ISO (2003–2008). ISO 13408 Aseptic Processing of Health Care Products Package. www.iso.org.

ISO (2003). ISO 13485:2003. Medical devices – quality management systems – requirements for regulatory purposes. www.iso.org.

Jacchetti E, Emilitri E, Rodighiero S, Indrieri M, Gianfelice A, Lenardi C, Podestà A, Ranucci E, Ferruti P and Milani P (2008). Biomimetic poly(amidoamine) hydrogels as synthetic materials for cell culture. *Journal of Nanobiotechnology*, **6**: 14 doi:10.1186/1477-3155-6-1.

Johnson PC, Bertram TA, Tawil B and Hellman KB. (2011). Hurdles in tissue engineering/regenerative medicine product commercialization: a survey of North American academia and industry. *Tissue Engineering Part A*, **17**(1–2): 5–15.

Kaplan DS, Hitchins VM, Vegella TJ, Malinauskas RA, Frondoza CG, Kimberly M, Ferlin KM and Fisher JP (2012). Centrifugation assay for measuring adhesion of serially-passaged bovine chondrocytes to polystyrene surfaces. *Tissue Engineering Part C: Methods*, **18**(7): 537–44.

Li Z, Ramay HR, Hauch KD, Xiao D and Zhang M (2005). Chitosan-alginate hybrid scaffolds for bone tissue engineering. *Biomaterials*, **26**(18): 3919–3928.

Liu H, Li X, Zhou G, Fan H and Fan Y (2011). Electrospun sulfated silk fibroin nanofibrous scaffolds for vascular tissue engineering. *Biomaterials*, **32**(15): 3784–3793. Epub 3 March 2011.

Mason C and Dunnill P (2008). A brief definition of regenerative medicine. *Regenerative Medicine*, **3**, 1.

Miller IJ (1996). Alginate composition of some New Zealand brown Seaweeds. *Phytochemistry*, **41**: 1315–1317

Messenger MP and Tomlins PE (2011). Regenerative medicine: a snapshot of the current regulatory environment and standards. *Advanced Materials*, **23**(12), H10–H17.

Mather ML, Morgan SP, White LJ, Tai H, Kockenberger W, Howdle SM, Shakesheff KM and Crowe JA (2008). Image-based characterization of foamed polymeric tissue scaffolds. *Biomedical Materials,* **3**: 015011–015022.

Mouriño V, Newby P and Boccaccini AR (2010). Preparation and characterization of gallium releasing 3-D alginate coated 45S5 Bioglass® based scaffolds for bone tissue engineering. *Advanced Engineering Materials*, **12**(7): B283–B291.

Otsu, N. (1979). A threshold selection method from gray-level histograms. *IEEE Transactions on Systems Man and Cybernetics,* **9**(1): 62–66. doi: 10.1109/TSMC.1979.4310076.

126 Standardisation in cell and tissue engineering

Pêgo AP, Siebum B, Van Luyn MJ, Gallego y Van Seijen XJ, Poot AA, Grijpma DW and Feijen J (2003). Preparation of degradable porous structures based on 1,3-trimethylene carbonate and D,L-lactide (co)polymers for heart tissue engineering. *Tissue Engineering,* **9**(5): 981–994.

Plagnol AC, Rowley E, Martin P and Livesey F (2009). Industry perceptions of barriers to commercialization of regenerative medicine products in the UK. *Regenerative Medicine,* **4**(4): 549–559.

RM 8396 – Tissue Engineering Reference, Scaffold available from www.nist.gov.

South GR (1979). Alginate levels in New Zealand Durvillaea (Phaeophyceae), with particular reference to age variations in D. Antarctica. Volume 9 i.e. Symp, (9): 133–142.

The Medicines for Human Use (Advanced Therapy Medicinal Products and Miscellaneous Amendments) Regulations (2010), http://www.legislation.gov.uk/uksi/2010/1882/schedules/made.

Torrisi L, Mondio G, Miracoli R and Torrisi A (2008). Laser and electron beams physical analyses applied to the comparison between two silver tetradrachm Greek coins. *35th EPS Conference on Plasma Phys. Hersonissos,* 9–13 June 2008 ECA, **32D**, 4.170.

Trieu HH and Qutubuddin S (1994). Polyvinyl alcohol hydrogels I. Microscopic structure by freeze-etching and critical point drying techniques. *Colloid and Polymer Science,* **272**(3): 301–309, DOI: 10.1007/BF00655501.

Wang CC, Yang KC, Lin KH, Liu YL, Liu HC and Lin FH (2012). Cartilage regeneration in SCID mice using a highly organized three-dimensional alginate scaffold. *Biomaterials,* **33**(1): 120–127. Epub 5 Oct 2011.

Wang Y, Tomlins PE, Coombes AGA and Rides M (2010). On the determination of Darcy permeability coefficients for a microporous tissue scaffold. *Tissue Engineering Part C,* **16**(2): 281–289.

Weir NA, Buchanan FJ, Orr JF, Farrar DF and Dickson GR (2004). Degradation of poly L lactide. Prat 2: Increased temperature accelerated degradation. *Proceedings of the Institution Mechanical Engineers, Engineering in Medicine,* **218**(H): 321.

Wu, L. and Ding, J. (2004). In vitro degradation of three-dimensional porous poly(DL lactide-co-glycolide) scaffolds for tissue engineering. *Biomaterials,* **25**: 5821–5830.

Yoon JJ and Park TG (2001). Degradation behaviors of biodegradable macroporous scaffolds prepared by gas foaming of effervescent salts. *Journal of Biomedical Materials Research,* **55**(3): 401–408.

© Woodhead Publishing Limited, 2013

7

Principles of good laboratory practice (GLP) for *in vitro* cell culture applications

B. IDOWU and L. DI SILVIO,
King's College London, Dental Institute, UK

DOI: 10.1533/9780857098726.2.127

Abstract: Good laboratory practice (GLP) is recommended in all leading laboratories as a framework of principles for performing cell culture studies. It provides guidelines for monitoring test systems, recording data, report writing and data archiving. GLP therefore assures regulatory authorities that any data submitted are indeed a true reflection of the results obtained during the study. Standard operating procedures (SOPs) should be in place containing comprehensive coverage of all critical phases of the study design, including management, monitoring and reporting.

This chapter reviews the principles of GLP for application to *in vitro* culturing of cells.

Key words: GLP, study director, cell culture, governing bodies, test system, quality assurance.

7.1   Introduction

This section describes in detail the history behind why GLP was set up, the role of GLP and areas where it can be used. Furthermore, it describes the laboratory which promoted the concept.

7.1.1   History of good laboratory practice (GLP)

The principle of good laboratory practice (GLP) was originally instituted in response to cases of animal fraud by pharmaceutical and industrial chemical (mainly pesticides) manufacturers in the latter part of 1970.[1]

In 1973, the Testing Laboratory Registration Act came into force in New Zealand; this defined the laboratory to include staff records, procedures, equipment and facilities. The Act also established a testing laboratory practice.[2] The formal concept of GLP, however, originated in the United States of America in 1978. It was during the early 1970s that the Food and Drug Administration (FDA) became aware of cases of poor laboratory practice and decided to investigate 98 laboratories, the majority in the USA, but some

127

© Woodhead Publishing Limited, 2013

128   Standardisation in cell and tissue engineering

in Europe. Fraudulent activities and poor laboratory practices were uncovered; examples included equipment which had not been calibrated to standard form and was therefore giving false measurements, incorrect and inaccurate accounts of the actual lab study, and inadequate test systems. One laboratory which was prominently noted during the investigation was the Industrial Bio Test (IBT) laboratory,[3] which was the largest contract-testing laboratory in the world and had conducted many hundreds of pre-clinical safety studies, mostly for the pesticide industry, over a long period. Up to 20000 studies had been conducted for hundreds of drugs and pesticides, but numerous irregularities were discovered in the data, including falsification of laboratory work. One of the companies that performed testing was Proctor and Gamble (P&G) acronym. In another example, in some of the tests performed by P&G, mice used to test cosmetics had developed cancer and died, but results were reported as good.[4] Non-US companies wishing to do business with the US had to comply with the US GLP regulations, and as a consequence different countries produced GLP regulations. In 1987, this established the FDA as the first government agency to assess laboratory compliance with GLP regulations. The FDA's action stimulated much interest in the US Environmental Protection Agency (EPA) and in other countries and organizations. In 1981, the Organization for Economic Co-operation and Development (OECD) produced GLP principles that are now international standard.[5,6]

7.1.2   Defining GLP

The FDA's Code of Federal Regulations (21CFR Part 58) defines GLP: 'A testing facility shall have Standard Operating Procedures (SOPs) in writing setting forth non-clinical laboratory study method that management is satisfied are adequate to insure the quality and integrity of the data generated in the course of a study'.[5] SOPs refer to a set of instructions for completing a specific task or operating a specific piece of equipment that has completed rigorous review and has been signed by the author, supervisor, documentation co-ordinator, quality assurance (QA) manager and facility reviewer. GLP aims to ensure good organisation of studies through a set of principles and quality systems that provide a framework within which laboratory studies, specifically non-clinical research, are controlled. 'The research must be planned, performed, monitored, reported and archived, ensuring uniformity, consistency, reliability and reproducibility. It must not be confused with the standards of laboratory safety like wearing a laboratory coat and gloves when conducting experiments'.[7]

In the European Union and other parts of the world, such as Australia, Brazil, Canada, Japan, Korea, Singapore and the US, it is a regulatory requirement that studies undertaken to demonstrate the health or environment safety of new biological or chemical substances should be conducted in compliance with GLP principles.[8]

7.1.3 Industries that use GLP

GLP is used in several areas of industries such as agrochemicals, cosmetics, veterinary medicines, industrial chemicals, food additives, contaminants and detergents, and for testing materials for space missions and experimental medical devices. GLP helps assure regulatory authorities that the data submitted are representative and reproducible, conform to the required regulations and operating procedures, and can therefore be relied upon when making risk/safety assessments. Complying with GLP can increase a laboratory's costs by 30%.[9]

A test facility can conduct both regulatory and non-regulatory studies. Briefly, regulatory studies are intended for review by the regulatory authorities, whilst non-regulatory studies are not intended for submission to the regulatory authorities. Non-regulatory studies are subjected to reduced QA monitoring and can therefore involve less specific GLP documentation.

Although the GLP regulations pertain specifically to non-clinical laboratory safety studies, they can be readily applicable to experimental research and development. The implementation of such standards in the research and development of biological agents for gene therapy would considerably enhance the existing measures employed to ensure product safety.[10]

Good laboratory practice (GLP), good clinical practice (GCP), good manufacturing practice (GMP) and others are regulatory requirements for the manufacturing and testing of drugs, biological products, human tissues and devices. GLP is considered the basic standard for non-clinical tests, whilst GCP must be in place for clinical trials and GMP for manufacturing cell therapy in patients, which therefore requires the most stringent regulations.

7.2 GLP governing bodies

Governing bodies exist in different countries, including Canada, Australia, the US and Europe. Within this section reference will be made mainly to the European governance, with some mention of the governing body in the US.

7.2.1 E-governance

Governance has been defined as 'the persons who make up a body for the purpose of administering something'.[11] E-governance, or electronic governance, uses new information and government technologies to make governance 'more efficient and more effective'. E-governance may thus improve service provision by improving communication between the government and its stakeholders, leading to a more inclusive and holistic method of operation.[12]

130   Standardisation in cell and tissue engineering

E-governance is based upon and strongly linked to democratic values and practices. E-democracy in turn necessitates e-governance. Numerous concepts, policies and practices of e-democracy are equally applicable to e-governance.[13]

The development and dissemination of common guidelines for good e-governance practice are essential to the sharing of expert knowledge across borders, learning from and building on the successes and difficulties of partners and avoiding duplication of work and the inefficient use of resources.

7.2.2 European Centre for the Validation of Alternative Methods (ECVAM)

ECVAM's main goal as defined by its Scientific Advisory Board is to promote the scientific and regulatory acceptance of alternative methods which are of importance to the biosciences and which reduce, refine or replace the use of laboratory animals.[14]

ECVAM and the Institute for *In Vitro* Sciences (IIVS) have put together a programme to enhance the principles of GLP. It drew on successful approaches used by industry, QA professionals and regulatory agencies with *in vitro* bioassays, by identifying additions and modification to the OECD's GLP principles to address the specific needs of *in vitro* methods (for example, test system characterisation facilities), and to formalise standards of practice to ensure quality data from *in vitro* studies.[14]

In 2011, the European Union Reference Laboratory for Alternatives to Animal Testing (EURL ECVAM) was established, due to the increasing need for new methods to be developed and proposed for validation in the European Union.[15]

EURL ECVAM has a long tradition in the validation of methods which 'reduce, refine and replace' the use of animals for safety testing and efficacy/potency testing of chemicals, biological and vaccines. It also promotes the development and dissemination of alternative methods and approaches. The key areas in which EURL ECVAM is active include topical toxicity and systemic toxicity, also covering biological agents which include a wide range of medicinal products such as vaccines, blood and blood components, somatic cells, gene therapy, tissues and recombinant therapeutic proteins created by biological processes.[16]

7.2.3 Interagency Coordinating Committee on the Validation of Alternative Methods (ICCVAM)

ICCVAM is an interagency committee composed of representatives from 15 US federal regulatory and research agencies that require, use, generate

© Woodhead Publishing Limited, 2013

Principles of GLP for *in vitro* cell culture applications   131

or disseminate toxicological and safety testing information to determine the safety or potential adverse health effects of chemicals and products to which workers and consumers may be exposed.[17,18]

The aim is to increase the efficiency and effectiveness of the federal test method review by eliminating unnecessary duplication and sharing experiences. It optimizes utilisation of scientific expertise outside the federal government and ensures that new and revised methods are validated to meet the needs of federal agencies and to reduce, refine or replace the use of animals in testing where feasible, in a similar way to the principles of EURL ECVAM.[17]

ICCVAM's duties also include the technical evaluation of new and alternative test methods, and the development of recommended tests based on technical evaluation; such information is then forwarded to federal agencies for their consideration. ICCVAM also coordinates interagency issues of toxicological test method development, validation, regulatory acceptance and national and international harmonisation.

National Toxicology Program (NTP) Interagency Centre for the Evaluation of Alternative Toxicological Methods (NICEATM)

NICEATM administers and provides scientific support for ICCVAM by co-ordinating and supporting meetings of ICCVAM and its subcommittees and interagency working groups, evaluating new test method submissions and nominations for their completeness and adherence to ICCVAM guidelines. They prepare technical review documents to evaluate new, revised and alternative methods and conduct independent validation studies on high priority alternative test methods.[17] ICCVAM then recommends appropriate uses of these methods to federal agencies, based on their scientific validity. Each federal agency then determines whether the test methods are acceptable for their respective programmes.

7.2.4 EU Tissues and Cells Directive

The EU Tissues and Cells Directive for regulation of cell-based therapies governs the use of tissues in research and clinical use, including safety and quality of cellular 'donation, procurement, testing, processing, preservation, storage and distribution'. The directive considers stem cell handling and processing when the cells are exposed to the environment provided by a Class 2 microbiological safety cabinet or a clean room facility.[12] These help govern cell usage in a way that makes the procedure safer and more cost- and time-efficient, and ensures a higher standard of practice. Where studies are for human application, systems are to be in place to ensure that all tissues and cells used are to be traceable from donor to recipient. In the

132   Standardisation in cell and tissue engineering

UK, the Human Tissue Authority (HTA) ensures that human tissue is used safely and ethically and most important with consent from the patient. This is strongly regulated and monitored.[19]

7.3   Resources required for GLP compliance

This section describes the resources required to be in place for the laboratory to be GLP compliant, for example: organization, personnel and management. Furthermore, the role of the study director and facilities and equipment used are also reviewed.

7.3.1 An organised structure

A laboratory seeking GLP-compliant status is required to have an organised structure in place. GLP requires the structure of the research organisation and the responsibilities of the research personnel to be clearly defined; the organisation chart should be kept up to date and reflect changes within departments. The chart should indicate where overall responsibility for GLP compliance lies within management and should include the lines of reporting and communication in the test facility. It should also include the main jobs areas associated with GLP and should identify the individuals who have designated responsibilities with respect to GLP organisation. This includes the maintenance of lists of individuals who can be appointed as study directors and principal investigators. The remaining positions include laboratory manager and test substance controller, SOP administrator, archivist for raw data and final report and QA personnel.[1,5,20,21]

7.3.2 Personnel and personnel management

GLP emphasises that the number of personnel available should be sufficient to perform the tasks required in a timely and GLP-compliant way. The responsibilities of all personnel should be defined and recorded within job descriptions, and their qualifications, together with a brief account of their education, employment history and professional experience both within the institution and before joining it should be recorded in the form of a CV. Publications, membership of associations and any further information pertinent to the position should be included. Personnel must clearly understand the functions they are to perform and where necessary be trained. Records should be kept which list specific tasks required to be undertaken. Job descriptions are to be presented in a GLP-compliant manner: job title, qualification and experience required, key objectives of the position, hours expected to be spent per week or day as appropriate, and clear definition of tasks and responsibilities. A review of all job descriptions should be

Principles of GLP for *in vitro* cell culture applications   133

undertaken annually or in the event of any reorganisation. All tasks to be performed without supervision should be formally allocated to the appropriate personnel. On completion the work is signed off; the worker thus takes full responsibility for its correct completion. Records of formal training showing the procedures which an individual can perform should form part of GLP documentation. Training records to a standard format should be available for all grades of professional staff with GLP duties. An SOP should describe who should have training records, who can authorise them and where they are to be located, and should rule that a training record should be archived when an individual no longer has GLP responsibilities. A file should be maintained for each member of staff containing all documentation including CV and training certificates.[1,5]

7.3.3 The study director

The role of the study director, who is a scientist or other professional of appropriate education, training and experience, is to oversee the elements of the work within the plan and to ensure the accuracy and validity of the studies performed. The study director should prepare a final report which accurately reflects the raw data generated during the study. These data should be fully and accurately documented and any amendments to the protocol should be reflected as appropriate. Study directors are also responsible for ensuring that adequate resources, SOP documentation and suitable trained personnel are available prior to commencement of the study. This includes approving the protocol by dated signature and ensuring appropriate liaison with and awareness of the protocol by all key staff involved in the study. The study director's role includes QA overall technical responsiblity; and regular communication with all personnel on the study including administrative, scientific and QA staff, to ensure that procedures are being effectively carried out.[1,8]

7.3.4 Facilities and equipment

The test facilities, including size and location, will depend on the test systems to be used and studies to be performed. Facilities include individual rooms and equipment provided to maintain the specified controlled environment for the test system(s). These requirements are necessary to ensure the study is not compromised due to inadequate facilities. Separation of cell lines where appropriate is necessary to ensure that different activities on the cells do not interfere, for example, that expansion of a range of cell types does not cause contamination. GLP facilities principles were originally developed in terms of rooms or areas, since such units are appropriate

134 Standardisation in cell and tissue engineering

for most *in vivo* studies where test systems are exposed to room air. In contrast, however, most *in vitro* systems such as cell culture are manipulated in vertical Class 2 microbiological safety cabinets to protect both the test system and the operator. The test system, that is, the cells, is expanded from tightly sealed membrane filtered flasks, or plates. Organisational separation should be arranged by carrying out different activities in the same area but at different times, such as allowing for cleaning and preparation between operations, and establishing defined work areas within the tissue culture laboratory.[1,5,8]

We can illustrate this by considering tissue culture rooms organised for different cell types and treatments. For example, some cells have been infected with a virus in order to immortalise cells; these will automatically have their own Class 2 microbiological safety cabinet and be kept in a separate room where non-immortalised cells are being grown. Animal cells and human cells should be kept separately in Class 2 microbiological safety cabinets and incubators to prevent contamination. Similarly genetic modification of cells by genetic engineering including gene targeting, which knocks out specific genes via engineered nucleases, necessitates separate cabinets.

Adequate storage facilities for consumables, culture ware and equipment, wash rooms, and autoclave facilities for autoclaving reagents are also required. There should be sufficient storage for samples/reagents in both refrigerators and freezers at $-20°$ and $-80°$. For cryopreserving cells a fortnightly supply of liquid nitrogen is required. Waste disposal should be conducted under safety conditions. Sharps bins are to be provided for correct disposal of needles, etc. Disinfectants must be available for the disposal of tissue culture waste and animal or human waste, and facilities must be in place to swiftly remove contaminants in culture.

Proper conduct of the study requires a supply of properly calibrated equipment of adequate capacity, which should be frequently cleaned and decontaminated. For example, the cell counting machine should be calibrated regularly and be maintained to ensure reliable and accurate performance for cell counting. Records of repairs and routine maintenance should be retained. The purpose of GLP requirements is to ensure that data are not invalidated or lost as a result of inaccurate, inadequate or faulty equipment. An SOP is therefore required for inspection, cleaning, maintenance and calibration of apparatus. This conformity is to be documented as raw data. The incubator, for example, should conform to the temperature range and ideally a record should be maintained, similar to that for fridges and freezers. Calibration and maintenance of all equipment is imperative to ensure accurate performance. For example, when several reagents are weighed on a balance, the accuracy of the balance must be maintained, using standard weights where appropriate. In the case of a pH meter, the use of standard chemicals is required to calibrate it and any other analytical equipment. Pipettes catering for various

© Woodhead Publishing Limited, 2013

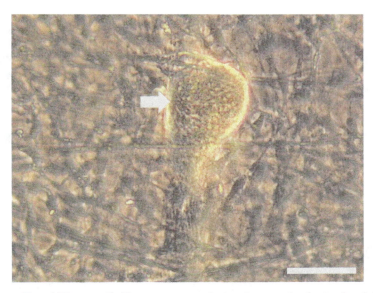

Plate I (Chapter 4) Phase contrast photomicrograph of rat calvarial osteoblasts cultured under osteogenic conditions for 21 days. Cells have begun to form multilayers and deposit mineralised matrix in a bone-like nodule (white arrow) (scale bar = 50 microns.)

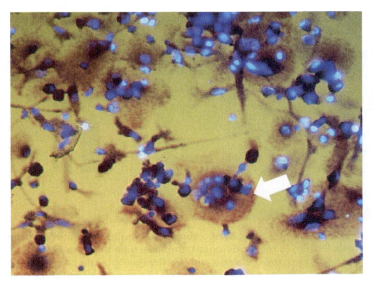

Plate II (Chapter 4) Photomicrograph of peripheral blood mononuclear cells cultured for 14 days under conditions that were supplemented with RANKL/M-CSF. Tartrate-resistant acid phosphatase (TRAP) staining (purple) and 4′,6-diamidino-2-phenylindole (DAPI) (blue) localisation illustrate the formation of multinucleate osteoclasts (white arrow).

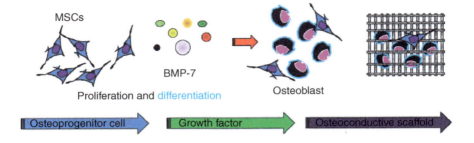

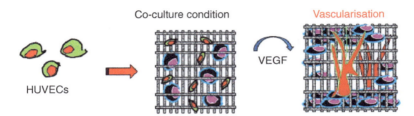

Plate III (Chapter 5) Co-culture protocol, mesenchymal stem cells (MSCs) seeded on scaffold with medium containing osteoinductive protein to commit cells toward osteogenic lineage; after osteogenic differentiation human umbilical vein endothelial cells (HUVECs) are added and the cell-scaffold continued to grow in co-culture media to allow *in vitro* prevascularisation.

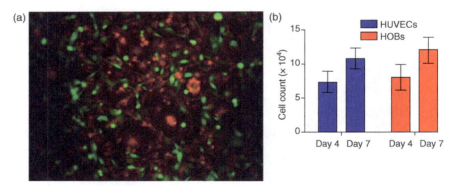

Plate IV (Chapter 5) (a) Image showing co-cultures of human osteoblast cells (HOBs) in red, and human umbilical vein endothelial cells (HUVECs) in green, stained with CellTracker™ (scale bar = 150 μm). (b) No significant difference between the numbers of each cell type at either day 4 or day 7, with p values of $p = 0.79$ and $p = 0.64$ for days 1 and 3, respectively.

© Woodhead Publishing Limited, 2013

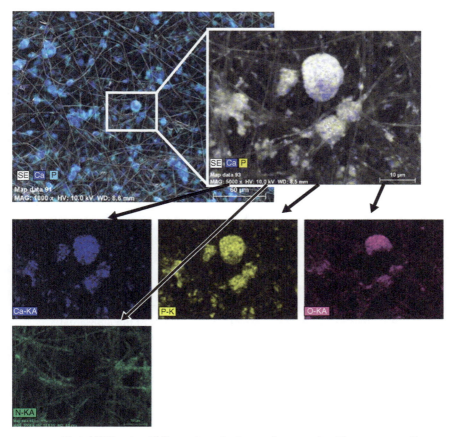

Plate V (Chapter 9) Scanning electron micrograph with corresponding elemental maps of calcium, phosphorus, oxygen and nitrogen. Image presents polymer scaffold with tri-calcium phosphate particles imbedded into scaffold structure.

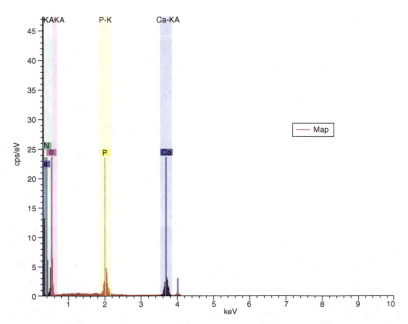

Plate VI (Chapter 9) EDS – quantitative analysis of the elemental composition of the electrospun PLA scaffold with calcium/phosphate glass filler.

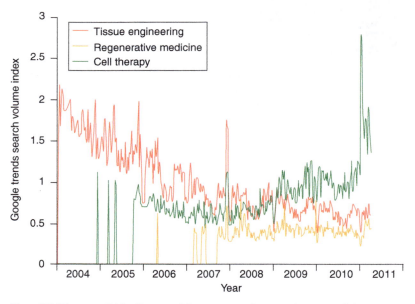

Plate VII (Chapter 11) Indicator of frequency of general usage of the terms 'tissue engineering', 'regenerative medicine' and 'cell therapy' (Mason *et al.*, 2011).

Principles of GLP for *in vitro* cell culture applications   135

volumes of liquid are in constant use and need to be calibrated annually. A liquid nitrogen storage alarm and an automatic back-up for fridge and freezer must be in place. Designated members of staff, preferably those able to respond quickly, should be on call after hours. All this, plus calibration protocols for all equipment, should be included in the SOPs.

Equipment used for capturing images, such as two-photon microscopes, and computer storage of images should be conducted and documented in a GLP-compliant manner. Before a computer is used for GLP studies it must be shown to be suitable for its intended use, with control systems in place providing for sufficient storage space and memory. A remote back-up system should be in place. The raw data for each computer system should be defined and the system design should always provide for the retention of audit trails where appropriate, to show all changes to data, without obscuring the original data. The focus of the acceptance testing effort should be on the application(s) being used, rather than on the computer system itself. It should be known beforehand whether the hardware and software together produce the desired result when challenged by rigorous testing with known data sets. Spreadsheets should be tested for their intended use and the performance documented. Similarly when images are captured on a two-photon microscope, the appropriate settings should, ideally, be known for all the different parameters.

Maintenance and calibration of all equipment should be documented and labels attached for quick visual inspection. Records of equipment calibration, checking and maintenance demonstrate that the respective SOPs have been followed.

Reagents, chemicals and solutions, including commercially obtained media and reagent solutions, must be labelled appropriately; labelling should be amended where appropriate, for example, supplementing the media with foetal calf serum or antibiotics or other labile components will affect expiry of the mixture, so it should be used within a short time to keep the additives effective. Additives should be stored appropriately and used within the recommended time period.

7.4   Characterisation

In this section the importance of documentation of data in order to satisfy the conditions of GLP status is described. The different test systems and cells used are summarized, and the relevance of controls and references are discussed.

7.4.1 Test system

A test system is broadly defined as a system exposed to a test item during a safety study. Test systems may be animals or plants, bacteria, organs, cells or analytical equipment. This chapter is concerned with cell and tissue culture systems.

Test system should be stored well before use (in liquid nitrogen/cryopreservation), handled carefully once removed and maintained under sterile conditions in order to ensure good quality data.

The cells or tissues used as test systems may be freshly isolated within the laboratory from human donors (called primary cells) or bought commercially, as either primary cells or established cell lines which have been serially propagated and maintained in continuous culture for long periods. They could also be an immortalised cell line. Maintenance and propagation of the test system must be addressed during the planning of a study and documented in sufficient detail. Characterisation of the *in vitro* system is of great importance and will be discussed in more detail.

Receipt of the test system, depending on the size of the organisation, may be delegated either to a central group, who will record the receipt of the test system, identify what it is and issue retention and final disposal documentation, where appropriate, of the test system, or to an authorised technician or the study director. In compliance with GLP, the assignment of responsibility should be documented in an SOP.

It is also important for the test system to arrive with the appropriate documentation, including source, passage history and manipulations. The media should include the raw materials of animal origin related to it. Information about exposure to contaminating agents and the degree of testing to determine freedom from contamination should also be included. All such information is required, since, for example, a primary/established cell line may not have arrived from the host site and therefore this information will need to be obtained for archive and store batch records.

Items transferred between facilities are to be examined, and packing is of great significance. Cells from commercial companies generally arrive in freezing medium or dry ice. Unexpected events, such as bad weather on delivery, should always be allowed for. The test system should arrive with the following information: company details or sponsor's name, date of despatch, number of vials, identity of test item, batch number, name of dispatcher and any information on testing for contamination with human blood-borne pathogens.

Proper storage, handling and care of test systems should be established on arrival to ensure quality of data. For example, if a vial of cells has arrived from a reputable company and has been allowed to sit in the lab for 24 h prior to unpacking, bearing in mind the instructions are to freeze on arrival, this will result in fewer viable cells, and as a result expansion of cells will take more time, increasing the passage time before use. Some cell types last up to only passage 8, for example, after which they stop dividing and senesce. On unpacking, high-quality cell and tissue culture practices must be enforced; this is an essential component of *in vitro* work.

Principles of GLP for *in vitro* cell culture applications 137

Before the test system is used, it is necessary to check that what has been sent is correct, and any significant changes will need to be evaluated.

Newly received test systems should be examined for purity and contamination; this can be monitored within 24 h after plating using aseptic techniques, i.e. under sterile conditions. Advanced testing is also recommended, entailing conducting routine assays for contamination with bacteria, fungi or mycoplasma. These tests should however, be conducted by a reputable commercial company. Alternatively a polymerase chain reaction (MycoTOOL®PCR mycoplasm Detection Kit from Roche) provides a sensitive and specific option for the detection of mycoplasm in cell culture. Until it is proved that the test sample is contamination-free and nothing in it can interfere with the purpose of the study, it should not be used in GLP-compliant studies and should be appropriately treated or destroyed. Records of the origin, maintenance requirements, source, date of arrival and arrival conditions are to be kept for tracking purposes.

After it has been established that the test system is contamination-free, cells should be allowed to propagate before use. An idea of the cell division should already be known, as demonstrated in the SOP and study plan, which may dictate that the cells reach over 80% confluence every 3 days and then split. Operators should prepare viable frozen cell stocks as soon as practicable; how early in the passaging it should be performed will vary depending on the cell type, although sooner rather than later is preferred. If the cells are demonstrating unexplained changes in growth rate or spontaneous differentiation it may be an indication of cell batch variation and this will need to be addressed prior to subsequent use, as it may well affect the biological characteristics of the cells.

All information needed to properly identify the test system, which in this case remains *in vitro,* should be adequately recorded throughout the course of the study. Individual test material that can be ordered in different volumes or sizes, such as flasks and plates, should be used from the same order number and batch where possible throughout the study. Sterile equipment such as forceps and pipette tips should be available and kept sterile for use at all times.

The maintenance of test systems *in vitro* in a stable undifferentiated state (depending on the cell type) remains a challenge for the operator. Whether it is intended for research or cell-based therapies, the culture of the test system will require standardisation. Key areas to be identified are: the definition of the test system; crucial quality criteria; ensuring maintenance of pluripotency (which is dependent on the cell type); examining proliferation capacity; and checking for microbial contamination. Standard differentiation criteria therefore need to be identified and defined. Appropriate training with SOP will need to be carried out and strictly followed to optimise replication and the chances of obtaining the same results where possible.

© Woodhead Publishing Limited, 2013

138   Standardisation in cell and tissue engineering

Depending on what the culture systems are to be used for, example, to induce differentiation, external factors will augment differentiation into different lineages. The use of soluble factors must be documented in the SOP and adhered to; quality and consistency should be monitored and any changes made must be added to the protocol. Since the batches of factors and serum might vary from batch to batch, screening of available batch must be included and documented. This may be time-consuming but is regulated to ensure a reliable and consistent test system.

Testing laboratories must use good scientific practices as well as appropriate calibration and standardisation methodology established by the various technical disciplines appropriate to the elements of the assay system. For example *in vitro* assays using cells in culture should follow good microbiological practices, for example, examining scaffolds which should show good biocompatibility. Scaffolds composed of materials derived from humans used in an *in vitro* model should ideally replicate an *in vivo* response.

If bone marrow has been obtained from patients to be used as a test system, a logging system should be set up; a unique identifier code will be supplied for each patient's bone marrow. A tracking system must also be able to track the tissue collected, its usage and storage. In the UK, such procedures should be facilitated and licensed under the Human Tissue Act (HTA), which was introduced in 2006 to regulate the storage and use of relevant human tissues such as bone marrow cells for use in research.

7.4.2   Test items including references and controls

GLP compliance requires test items in terms of reference and control to be in place. The control serves to monitor the performance of the test system, for example if the viability of the cells is being examined, one would have a control well with medium only or with a reagent that will kill the cells. Procedures designed to prevent errors in the identification and cross-contamination of the test, control and reference items and their respective preparations must be in place.

Records of these procedures should be supervised from the outset and be maintained throughout the course of the study until the final deposition, be it archiving, return or disposal.

7.5   Standards and regulations

In this section the design of the study is highlighted with reference to the approval requirements by the nominated person. The importance of maintaining a standard operating procedure, to ensure quality and integrity of data generated is discussed.

Principles of GLP for *in vitro* cell culture applications   139

7.5.1 The protocol

The content of the protocol must be coherent with the scientific requirements of the study and should also comply with GLP. The protocol describes the design of the study, including the overall time frame and an indication of the methods and materials to be employed. The study should be approved by the study director and verified by QA personnel. The protocol is the principal source of instruction for study staff about method, contents, style and layout of the study.

The protocol should have a study number or a code name providing a unique identifier for laboratory records which are connected to the study and confirming the identity of all data generated during the study. It should include negative and positive control items to be used as additional information where appropriate.

The protocol should contain a short but informative title and statement of purpose, specifying the type and duration of the study, the test system, and the reason why the study is being done. Scientific guidelines may be referred to where appropriate.

The sponsor or principal investigator is the person who initiates and supports the study by providing financial or other resources in a non-clinical laboratory, or who submits a non-clinical study to the regulatory body in support of the study. The protocol should indicate where the study will be performed and other organizations that may be associated with it. Where studies are to be conducted on multiple sites this has to be included in the study plan.

The name and address of the study director and other responsible personnel must be included in the protocol, including the principal investigator who will make a significant contribution to the study. If the study is to be conducted at multiple sites, a particular test site must be allocated to the principal investigator who will, therefore, be the person responsible for the test phase conducted at that site.

The proposed start and finish dates for the study noted in the protocol must correspond to the date the report has been signed by the study director. A detailed schedule is recommended to assist staff. Any change of date should be defined in the SOP for protocol management.

Reasons for the choice of test system must be given, including the characterisation of the *in vitro* system, such as the species and tissue of origin, source of supply, cell designation, culture conditions and other relevant information, which may well be dependent on the test facility.

Before a GLP study begins, the protocol should be planned and approved, shown to all staff concerned and its implications discussed, and circulated to QA for review and approval. A study initiation meeting should also be arranged. All staff in receipt of the protocol will need to sign a document to confirm they are fully aware of their individual roles.[1,5]

© Woodhead Publishing Limited, 2013

140 Standardisation in cell and tissue engineering

Any changes during the study period should be made in consultation with the study director. Changes to the study design must be justified and any modifications made using an agreed process known as a change control procedure. The important elements of a protocol amendment include: clear identification of the study being amended; clear new instructions; unique numbering of the amendment; clear identification of the section of the original protocol being amended; issue and clarification of the amended protocol to the same staff as the original protocol.

7.5.2 Standard operating procedures (SOPs)

A complete set of SOPs is to be in place for successful GLP compliance, and this is regarded as the most time-consuming of the compliance tasks. It should be approved by the test facility management, with evidence of their supported and commitment to its preparation. Deming and Juran suggest: 'Use standards (SOPs) as the liberator that relegates the problems that have already been solved to the field of routine, and leaves the creative faculties free for the problems that are still unsolved'.[22]

The SOPs are intended to ensure the quality and integrity of the data generated by the test facility. Any revision must be approved by the test facility management. SOP-based education and training of personnel should be provided, such that procedures are carried out in the same way by all involved.

Each separate test facility unit should have immediate access to current SOPs relevant to activities being performed. SOPs are not standalone and are to be supplemented with published text books, analytical methods, articles and manuals. SOPs should have the following characteristics: they should be integrated in the laboratory's system of master documents; and they should contain comprehensive coverage of all critical phases of study design, management, conduct, monitoring and reporting. A standard layout is required including title, purpose, background information on the test, instructions for the procedure in chronological order, references and a named contact person in case of emergency or query. A section heading and section numbering system should be adopted to promote consistency. All staff should be encouraged to contribute, thereby providing a variety of authors reflecting the expertise of many individuals within the organisation. It should be reviewed and approved by the appropriate signatory.

For reasons of traceability and ease of use, SOPs should consist of two sections, one containing general policies and procedures and the other describing technical methods. General polices includes protocol writing, review, approvals, distribution and modification, general rules for equipment use and maintenance, and data archiving, whilst technical methods would include, for example, methods of immunofluorescent staining. SOPs

© Woodhead Publishing Limited, 2013

Principles of GLP for *in vitro* cell culture applications   141

are usually provided as manuals with an up-to-date table of contents; some laboratories, however, have electronic SOPs that are password protected. It is important to prevent unauthorised access and designated individuals will be asked to monitor usage. All alterations to SOPs have to be made as formal revisions, notes and changes; they should not be hand written which may cause errors. It is imperative that all staff are fully conversant with the SOPs they use and adhere to them rigorously; this will ensure data reproducibility and validity. A designated person should be responsible for keeping SOPs updated and handling queries.[1,8,22,23]

New test methods and approaches to toxicity testing include high-throughput screening, non-mammalian animal systems, computational approaches, toxicity biomarkers and development of high-quality toxicology databases.

7.6   Documentation of results

This section describes the importance of documentation of results to satisfy the conditions of GLP status, including collection and storage of raw data and the reporting and archive of the data.

7.6.1 Raw data

Raw data are described in the ECVAM Workshop Report 37 as

> all original test facility records and documentation, or verified copies thereof which are the result of the original observations and activities in a study. Raw data can also include photographs, microfilm or microfiche copies, computer readable media, dictated observations, recorded data from automated instruments, or any other data storage medium that has been recognised as capable of providing secure storage of information for the required time period.[8]

Raw data should be recorded at the start and finish of the assay or test, and this is an essential part of the study. Data must include the name of the operator(s) and each person's responsibilities, and the date and where possible the time of the procedure. General information such as equipment, materials, reagents including concentrations and volume used, batch used, control and reference items, of test systems should be recorded. The detailed method should describe what was done, including the techniques used, the results of the observations, using images where possible to indicate all actions including the protocol used. It should describe in detail the methodology set out in the SOP; notes should be pencilled in if any modification was made during the course of the procedure. The data must be recorded accurately and legibly as soon as the procedure has been completed in such a way as to allow no doubt as to the credibility of the data. It

© Woodhead Publishing Limited, 2013

142 Standardisation in cell and tissue engineering

must be signed off, and dated, for accountability, which is one of the tenets of GLP. At the end the raw data are to be maintained in a safe storage area prior to long-term secure archiving.[1,8]

7.6.2 Final report and archiving

The final report must include the following: the name and address of the test facility; the name of the study director; a GLP compliance statement from the study director; the study objectives; details of the test systems including its objectives; the precise dates experimental work commenced and ended; and details of the equipment and reagents used. It should contain an account of the practical conduct of the study, and also demonstrate that the contents of the report describe the study accurately. It should clearly state if there were any changes in the procedure or SOP, and whether this is considered to have impacted on the study integrity or not. The results should be tabulated to show significant features, and should include statistical analysis and discussions with a list of references. The report should provide evidence of signed and dated reports from all the scientists responsible. There should be a critical discussion of the study, which should end with a conclusion. The report must be dated and signed by the study director to confirm responsibility for the validity of the data.

The report should then pass through a review stage and a QA audit, to ensure that it is unlikely to require further modification.

At the end of every study, all data from the study must be collected together with the study plan and final report and combined into a single package of information referred to as the study file. Good GLP requires it to be formally archived to guarantee the integrity of the data and hence the study. The organisation chart should include a designated archivist responsible for securing orderly storage, which is important for rapid retrieval of reports. Finally, access to the report should be restricted. It may be examined in the presence of the archivist, and photocopies may be supplied upon request.[1,8] National GLP regulations vary as to the retention period of the data, which may be between 10 and 15 years.[23–25]

7.7 Independent monitoring of research processes and quality assurance (QA) personnel

The person responsible for the QA of cell culture work is independent of the organisation and research process. The role of QA personnel role is to ensure that the experimental protocol complies with GLP principles. Before study commences they should review all phases of non-clinical studies, including a list of all studies planned. All SOPs should be reviewed frequently for clarity by the designated personnel. Inspection by QA personnel is ongoing during

© Woodhead Publishing Limited, 2013

Principles of GLP for *in vitro* cell culture applications   143

the study and includes: regularly monitoring facility operations; reviewing studies, documents and schedules; keeping appropriate records of audits and inspections; and reporting to management as appropriate. Inspections and audits performed in compliance with the QA programme should be described in an SOP. An audit or inspection is a methodological evaluation of a process, an activity or a set of data, but must not be considered as an inquisition process. QA personnel may sign the SOP document in order to acknowledge it has passed a GLP review, although this is not compulsory, and it does not interfere with the technical and scientific conduct of the studies. QA personnel should have access to all raw data either during inspection or in the final reports; where identical repeats have been conducted on experimental studies, there should be signs of reproducibility. QA personnel should monitor any action taken in response to the report. Any area of non-compliance that may have occurred during the study should be duly noted and explained, and of course must be included in the statement.

QA documents and procedures should be familiar to staff at all levels of the organisation, including management. Once the report has been signed by the study director, QA personnel may receive a copy of all protocols with any subsequent amendments. QA inspections are scheduled in advance with the study director and representative from the personnel.

A QA statement is generated which is included in the report with the date when study was audited and inspected, and findings are reported to the study director and management. The statement therefore provides a clear formal indication that GLP compliance has been followed and every reasonable step has been taken to generate an accurate quality report reflecting quality raw data generated during a study that has been conducted in compliance with GLP. The statement, however, is not an absolute guarantee of scientific excellence or quality.[1,8,20]

7.8   Application of GLP to human cell culture systems

This section describes in detail a specific human cell culture system. Examples are given of the procedure used for isolation of stem cells from bone marrow. The importance of correct storage and preserving the specimens is discussed.

7.8.1 Procedure for the isolation of stem cells from bone marrow

Prior to obtaining bone marrow cells from patients, consent must be obtained under the Human Tissue Act.[19] Figure 7.1 shows the procedure once this has been obtained. There are several methods for extracting stem

© Woodhead Publishing Limited, 2013

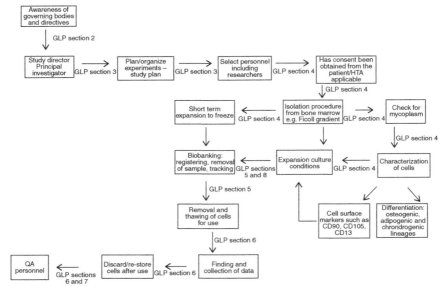

7.1 Schematic summary of events once applied to a cell culture system.

cells from bone marrow, such as the Ficoll-Paque technique for the isolation of mononucleated cells followed by the separation of MSC for adherence to plastic; Ficoll-Paque followed by the immune-magnetic separation of MSC; and finally a negative selection procedure of MSC using the RosetteSep technique, followed by adherence to plastic. The method of choice for this system was, however, Ficoll-Paque technique allowing for adherence to plastic. Briefly, the method involved diluting the bone marrow sample with cell culture medium supplemented with PBS and EDTA and pouring on the Ficoll-Paque solution in a 50 mL centrifuge tube. This was then gently centrifuged, producing an interface of cells consisting of mononucleated cells accumulated on the Ficoll–plasma interface. Centrifugation was repeated and cells were counted before seeding into flasks. Adherence cells were the stem cells, which were characterised as described in Fig. 7.1.

It is always preferable to preserve some cells at an early passage, hence the need to expand for a short time in culture, and then freeze down. Cells should be checked for mycoplasm (Fig. 7.1), but in order to ensure stem cells have been extracted, the cells would need to be characterised. This would involve expanding the cells further using the following methods: flow-activated cell sorting (FACS), or flow cytometry, using cell surface markers; the markers specific for mesenchymal stem cells are CD90, CD105 and CD13, whilst for haemopoietic stem cells the markers used are CD45, CD34, CD133 and VEGFR2. Subsequently cells would be differentiated into different lineages, including the commonly reported osteogenic, adipogenic, chondrogenic and tenocytic lineages. It is then suggested cells are expanded for long periods to determine if the batch of cells undergoes transformation.[26]

7.8.2 Biobanking and biopreservation

Biobanks, which may also be described as biological resource centres (BRCs), are resource culture collections. Biobanks are not exclusive to pharmaceutical companies, but may also be present in academic and routine testing laboratories, and may be public or private services.[27] It is necessary to ensure that the material is free from contamination on collection[28] to prevent erroneous data, and that it is authentic to ensure stable phenotype and genotype characteristics remain. This may be confirmed by DNA profiling.[29] It should also be ensured that the cell maintains its features after serial passages, and that it has not undergone irreversible changes as a result of being maintained in culture for long periods of time.[30] Finally, GLP requires any preserved cultures to have data demonstrating the value of the stored culture in terms of results of experiments carried out and/or published work.

7.9 Conclusion

GLP compliance lays down certain criteria for a tissue culture lab. This chapter has described the principles of an *in vitro* tests system using cells which may or may not be directly obtainable from a patient. If human tissue is used, the Human Tissue Authority (HTA) is required to be in place and be commissioned by the appropriate personnel. History describes lessons learnt from errors committed knowingly and unknowingly, but every effort has been made to avoid recurrences by putting guidelines in place through the relevant governance provisions. The study director who oversees the study plan should ensure an organised structure is in place, including having the appropriate protocols alongside the SOP. The facilities, equipment, test systems and controls, documentation of data, archiving and finally QA personnel should all follow GLP to ensure a fully compliant GLP laboratory for cell culture.

7.10 Acknowledgements

This chapter presents independent research commissioned by the National Institute for Health Research (NIHR) under the i4i programme and the Comprehensive Biomedical Research Centre at Guy's & St Thomas' NHS Trust. The views expressed in this publication are those of the author(s) and not necessarily those of the NHS, the NIHR or the Department of Health.

146 Standardisation in cell and tissue engineering

The authors also acknowledge support from the Centre of Excellence in Medical Engineering funded by the Wellcome Trust.

7.11 References

1. Ridley R. (2009) *Handbook. Good Laboratory Practice (GLP): Quality Practices for Regulated Non-Clinical Research and Development*, 2nd edn. Switzerland, World Health Organization, 9–57.
2. New Zealand Legislation http://www.legislation.govt.nz/act/public/1972/0036/latest/whole.html.
3. Department of Health, Education and Welfare, Food and Drug Administration. Nonclinical laboratory studies: Proposed regulations for good laboratory practice 1976. *Fed Reg* **41**, 51205–51230.
4. Schneider K. (1983 (Spring)). Faking it: the case against industrial bio-test laboratories. *Amicus J* (Natural Resources Defence Council), 14–26.
5. Good Laboratory Practice: Code of Federal Regulations 21. Available from: www.accessdata.fda.gov/scripts/cdrh/cfdocs/cfcfr/CFRSearch.cfm?CFRPart=58&showFR=1 (Accessed August 2012).
6. Ertz K. and Preu M. (2008). International GLP: a critical reflection on the harmonized global GLP standard from a test facility viewpoint. *Ann Ist Super Sanità* **44**(4), 390–394
7. Medicine and Healthcare Regulatory Agency: Definition of Good Laboratory Practice (2006). Available from: www.mhra.gov.uk (Accessed August 2012).
8. National Websites for Good Laboratory Practices: Available from: http://www.oecd.org/env/chemicalsafetyandbiosafety/testingofchemicals/linkstonational-websitesongoodlaboratorypractice.html (Accessed August 2012).
9. Cooper-Hannan R., Harbell J.W., Coecke S., Balls M., Bowe G., Cervinka M., Clothier R., Hermann F., Klahn L.K., de Lange J., Lievsch M. and Vanparys P. (1999). The principles of Good Laboratory Practice: application to *in vitro* toxicology studies. The report and recommendations of ECVAM workshop 37. *Altern Lab Animal* **27**, 539–577.
10. Medicine and Healthcare Regulatory Agency – Regulatory Guidelines (2009). Available from: www.mhra.gov.uk/Howweregulate/Medicines/Inspectionandstandards/GoodClinicalPractice/index.htm (Accessed August 2012).
11. Simpson JW. (2008). *Oxford English Dictionary*. Oxford: Oxford University Press
12. Grange S. (2011). Tissue engineering stem cells – an e-Governance Strategy. *Open Orthopaed J*, **5**, 276–282 (Suppl2-M8).
13. Uses of E-governance – Available from: http://www.coe.int/t/dgap/democracy/activities/GGIS/E-governance/ (Accessed August 2012).
14. OECD (1998). Organisation for Economic Co-operation and Development series on Principles of Good Laboratory Practice and compliance monitoring. Number 1. OECD principles on Good Laboratory Practice (as revised in 1997), OECD Environmental Health and Safety Publications. Environment Directorate: ENV/MC/CHEM(98)17. Paris, France: OECD.
15. EURL ECVAM Alternatives to animal testing. Available from: http://ihcp.jrc.ec.europa.eu/our_labs/eurl-ecvam (Accessed August 2012).

Principles of GLP for *in vitro* cell culture applications   147

16. EURL ECVAM Alternatives to animal testing. Available from: http://ecvam.jrc.it/ (Accessed August 2012).
17. Interagency Co-ordinating Committee on the Validation of Alternative Methods. Available from: http://iccvam.niehs.nih.gov (Accessed August 2012).
18. ICCVAM Federal Government. Available from: http://iccvam.niehs.nih.gov/docs/about_docs/PL106545.pdf Public Law 106-545 Dec 2000 (Accessed August 2012).
19. Human Tissue Authority. Human Tissue Act (2004). Available from: http://www.hta.gov.uk/legislationpoliciesandcodesofpractice/legislation/humantissueact.cfm (Accessed August 2012).
20. Good Laboratory practices Guidelines. Available from: http://www.sjsu.edu/faculty/chem55/55glpout.htm (Accessed August 2012).
21. Coecke S., Balls M., Bowe G., Davis J., Gstraunthaler G., Hartung T., Hay R., Merten O., Price A., Schechtman L., Stacey G. and Stokes W. (2005). Guidance on Good Cell Culture Practice. A report of the second ECVAM task force on good cell culture practice. *ATLA* **33**, 261–287.
22. Zhou M. (2011). *Regulated Bioanalytical Laboratories. Technical and Regulatory Aspects from Global Perspectives.* USA: John Wiley & Sons, pp. 1–13.
23. Froud S.J. and Luker J. (1994). Good Laboratory Practice in the cell culture laboratory. In *Basic Cell Culture: A Practical Approach* (ed. J.M. Davies), UK: Oxford University Press, 273–286.
24. Organisation for Economic Co-operation and Development. *Consensus document of the Working Group on GLP: The application of the OECD principles of GLP to the organisation and management of multi-site studies.* Paris: OECD; June 25,2002. (OECD Series on Principles of GLP and Compliance Monitoring, No. 13, ENV/JM/MONO(2002)9).
25. Weinberg S. (2007). *Good Laboratory Practice Regulated.* 3rd Edn. Revised and Expanded. Informa Healthcare NY, Marcel Dekker **168**, 151–166.
26. Bernardo M.S., Zaffaroni N., Novara F., Cometa A.M., Avanzini M.A., Moretta A., Daniela Montagna D., Maccario R., Villa R., Daidone M.G., Zuffardi O. and Locatelli F. (2007). Human bone marrow–derived mesenchymal stem cells do not undergo transformation after long-term in vitro culture and do not exhibit telomere maintenance mechanisms. *Cancer Res* **67**, 9142–9149.
27. Day J.G. and Stacey G.N. (2008). Biobanking. *Mol Biotechnol* **40**, 202–213.
28. Mutewa S.M. and James E.R. (1984). Cryopreservation of *Plasmodium chabaudi* II. Cooling and warming rates. *Cryobiology,* **21**, 552–558.
29. Stacey G.N., Byrne E. and Hawkins J.R. (2007). DNA fingerprinting and characterization of animal cell lines. In R. Poertner (ed.), *Animal Cell Biotechnology: Methods and Protocols Totowa*, 2nd edn, NJ: Humana Press pp.123–145.
30. Stacey G.N. (2002). Standardisation of cell lines. *Dev Biol* **111**, 259–272.

8
Quality control in cell and tissue engineering

I. B. WALL, University College London, UK and
N. DAVIE, University of Oxford, UK

DOI: 10.1533/9780857098726.2.148

Abstract: Quality control (QC) and quality assurance (QA) are
fundamentally important for commercial manufacture of medicines,
particularly so for delivery of cells for therapy. Continued assessment of cell
product quality throughout the manufacturing process, either directly or
by developing highly defined processes that meet stringent QA standards,
is essential to ensure that the products are of a high quality and that high
reproducibility between batches is achievable. This chapter addresses some
of the issues that have to be considered for ensuring that cell products
delivered to the patient are safe and effective. Crucially, considerations for
the safety, purity, identity and potency of the cells are discussed.

Key words: bioprocessing, biomanufacturing, commercialisation,
translation, cell therapy manufacture, tissue engineering, process
development, critical quality attributes, quality control, quality assurance,
autologous.

8.1   Introduction

The concepts of engineering new body parts in the laboratory to replace
worn or damaged tissue or delivery of cell therapies to cure disease are
exciting.

8.1.1 Quality control (QC) in cell therapy manufacture

'Quality' has been a key assessment criterion of medicinal products since their
inception. Over time, treatments have become more refined and the standard
of product quality expected by both governing bodies and patients has grown.
The evolution of hundreds of years of rising standards has culminated in two
main methods of ensuring excellence: quality control (QC) and quality assurance (QA). These two often confused assessment methods have hugely different implications in terms of time, process and financial requirements.

148

© Woodhead Publishing Limited, 2013

Quality control in cell and tissue engineering   149

- Quality control is a set of measures implicated by the manufacturer at various points throughout a process to ensure that the product is of an acceptable standard (e.g., inspection, testing).
- Quality assurance refers to the process used to create the product, as well as alternative methods to ensure product quality, such as standardisation in staff training.

QC and QA are vitally important in the large-scale production of cell-based therapies. As cells are living units, they are fragile, extremely sensitive to their environment and can be prone to apoptosis if improperly handled. Purifying cell populations can also be challenging, especially where the end product is a mixed population or not easily characterised by cell surface markers. Consequently, the mantra 'the product is the process' becomes very significant (Mason and Hoare, 2007). Robust validation and assessment of the process against QA criteria will ensure that the input cells have undergone highly standardised and reproducible bioprocessing steps.

Furthermore, the 'whole bioprocess' is not limited to the cell culture phase but encompasses all the steps from initial tissue biopsy from the patient, cell liberation by tissue digestion, any expansion, differentiation and seeding onto biomaterial scaffolds that might be required, right through to transplantation into the patient. Any stage of the whole bioprocess, if not carefully controlled, could invoke a phenotypic drift in the cell population (Carmen *et al.*, 2012). It is crucial that by the end of bioprocessing, cells are phenotypically stable and capable of effectively doing the job they were made for every time and for every patient.

A particular issue for autologous cells relates to the inherent variability between samples, which can be hugely significant, making cells incomparable with other therapeutics such as small molecule pharmaceuticals (Mason and Dunnill, 2007; Mason and Hoare, 2007). Existing QC procedures for conventional products are, further, inapplicable, as cell therapy products require more sensitive and specific methods for QC, as often only low numbers of cells (approximately 1×10^6–1×10^8) are available for testing (Hashii *et al.*, 2007). For these reasons, a robust quality management system is paramount to having a successful bioprocess that satisfies regulators.

There are several key regulatory bodies who apply guidance and legislation for the cell therapy industry. On a global scale, the International Conference on Harmonisation of Technical Requirements for Registration of Pharmaceuticals for Human Use (ICH) works to resolve discrepancies between regulators in Europe (European Medicines Agency), Japan and the US (Food and Drug Administration, FDA). The differences between these governing bodies, although notable, are outside the remit of this

© Woodhead Publishing Limited, 2013

book chapter. Rather, our main focus will be the QC aspects of cell therapy bioprocessing.

8.1.2 Quality control requirements for allogeneic vs autologous cell-based therapies

Production of both autologous and allogeneic cell therapy products have similar processing requirements, in spite of the different achievable manufacturing scale of each. In fact, in terms of QC the main difference between autologous and allogeneic products is the cost. In Fig. 8.1, a schematic representation of relative cell quantities required for QC testing in autologous and allogeneic processing is shown, reproduced from Brandenberger *et al.* (2011). The fundamental difference is related to batch size. In an allogeneic process, one batch is used to treat several different patients and so the cell number required for testing represents a small proportion of the total cell production. Conversely, in autologous processes, one batch makes enough cells for one patient and therefore a certain degree of overproduction is necessary to account for the cells required for testing. More significantly, the volumes of reagents used in QC assays increases exponentially and consequently QC contributes to a large proportion of the total cost of an autologous product (Brandenberger *et al.*, 2011). Some allogeneic therapies fall under this umbrella too, such as donor-matched cord blood administrations, where processing is essentially for an allogeneic therapy but produced on an autologous scale, as one donor is required to treat one patient.

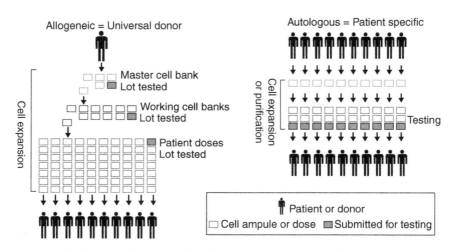

8.1 Testing challenges in cell therapy manufacture (Brandenberger *et al.*, 2011). (This figure is reprinted with permission from *BioProcess International*, March 2011, pages 30–37.)

8.2 Quality control to ensure a well-defined cell therapy product

The emerging cell therapy and tissue engineering industries are experiencing various 'growing pains' when it comes to standardising regulation and quality management. Translating research conducted in the laboratory into an industrial process requires protocol standardisation to attain the levels of consistency that regulators demand and scalability needed to meet the clinical demand and ensure the best chances of commercial success (Brandenberger *et al.*, 2011). These requirements, as well as the subsequent comparability studies, are expensive in terms of both time and money. Each stage of manufacture requires rigorous, robust and reproducible testing to ensure that the product is of a pre-defined quality.

In order to assess product quality, it is necessary to understand the fundamental features of the cell or tissue-engineered product by means of characterisation. However, a fundamental problem, particularly for cell-based therapies, is that very often the exact nature of the transplanted cells and their mode of action is undefined, and as a result the *product is the process* by which the cell therapy was derived (Mason and Hoare, 2007) and therefore the process has to be highly defined and operate within a tightly controlled set of parameters. This is particularly the case for autologous products, where it often isn't feasible to test the final product due to the limited quantity of material available for transplant.

8.2.1 Defining critical quality attributes for cell therapies

Cell product characterisation needs to take into account four important aspects:

1. Safety, to ensure that the material being transplanted is compatible with the recipient, is free from contaminating microbes and does not carry any genetic instability.
2. Purity, to ensure that the appropriate cell population is being transplanted and is not contaminated by unidentified or, in the case of pluripotent stem cell-derived products, undifferentiated cells that might cause harm.
3. Identity, to ensure that cells with the appropriate phenotype to carry out the proposed mode of action are transplanted.
4. Potency, to ensure that an acceptable level of clinical efficacy of the therapy is achieved.

These areas have been discussed at length (Carmen *et al.*, 2012) and are fundamental to good quality management.

152 Standardisation in cell and tissue engineering

Safety

During cell therapy manufacturing, FDA guidelines give safety characterisation the highest priority (Carmen *et al.*, 2012). The purpose of safety characterisation is to ensure that the end product is sterile, uncontaminated by bacteria, adventitious viral agents or mycoplasma and that, where applicable, the cells do not carry genetic instability that could cause tumourigenesis or malignancy in the patient. Sterility testing is a very important step in preparation of cell-based therapeutics (Goldring *et al.*, 2011).

Contamination by microbes is a common concern for cell therapy and tissue engineering protocols, as to date, many of the manufacturing methods are still heavily reliant upon manual processing and the operator is regarded as the most probable source of contamination throughout the whole bioprocess (Mason and Hoare, 2007). Additional risk of contamination from culture medium and other input materials and cell source itself (particularly in the case of autologous therapy where there is risk of contamination at biopsy) are all important considerations for the production of a cell-based or tissue-engineered therapeutic (Ratcliffe *et al.*, 2011). Use of antibiotics should be avoided as it can mask the presence of microbial organisms, which then become detectable upon removal of the antibiotic, or worse still, initiate pathogenic events upon transplantation. Cells should be tested for the presence of bacteria, fungus and mycoplasma routinely in any cell culture lab (Stacey and Stacey, 2000), but adherence to strict testing protocols becomes even more important when working with cells that might be used for therapy, and microbial detection tests should be conducted after a minimum of 5 days' culture and preferably after two antibiotic-free passages to ensure that any suppressed microorganisms do not go undetected (Stacey and Auerbach, 2007). Microbial testing kits are therefore routinely used to test for microbial presence as a primary QC measure.

Conventional antimicrobial testing for cell therapy manufacture, adhering to the standard 14-day sterility panel prescribed by FDA regulation 21 CFR 610.12 (FDA, 2011) can be problematic for cell products where the shelf life is very short, sample quantity for therapy is very small, or where the clinical indication requires cells to be delivered to the patient within a few days. As a result, cells are often delivered to patients before the results of standard microbial testing have been obtained. In fact, in the case of one current cell-based therapy for the treatment of advanced prostate cancer (see Chapter 11), patients sign a disclaimer acknowledging that test results will not be ready prior to cell delivery (Higano *et al.*, 2010). Therefore draft Guidance for the Use of Rapid Microbiological Methods has been introduced by the FDA to facilitate development and validation of assays that provide equivalent data to the sterility panel test in a shorter time-frame (FDA, 2008a). A notable example of the introduction of rapid

Quality control in cell and tissue engineering   153

microbiological methods in industry is Genzyme, who developed and validated BacT/ALERT, a method to reduce the time for detection of microbial contaminants to just 48 h for most organisms (Kielpinski *et al.*, 2005), signifying a substantial reduction in detection time. Even more rapid detection can potentially be achieved using molecular methods that reduce detection time considerably – these techniques are capable of amplifying even minute quantities of pathogen DNA to detectable levels for identification (Rayment and Williams, 2010; Sampath *et al.*, 2010). In an industry where the shelf life of a product can be as short as a few days, rapid sterility testing is hugely advantageous.

In addition to microbial contamination, viruses can remain dormant for protracted periods of time, making the accurate detection of viral contamination both challenging and incredibly important. Although donor material may be a source of viral contamination, the donor is screened for transmittable diseases including hepatitis B and C, human T-lymphotrophic virus, HIV and CJD (FDA, 2008b). However, it is not just human viruses from donor material that can be pathogenic – many culture reagents still contain animal products such as foetal calf serum (FCS), which is a potential source of contamination, particularly by mycoplasma (FDA, 2008b; Carmen *et al.*, 2012). The availability of high-sensitivity mycoplasma detection kits that provide data very rapidly (within one hour 1 h) provides a good QC tool for screening (Rohde *et al.*, 2008).

Use of FCS in cell culture for cell or tissue engineering for transplant into patients has also been questioned due to the risk of transmitting viral elements or prions that cause serious illness such as those responsible for transmissible spongiform encephalopathies (TSEs). However, there is a misconception that regulatory requirements stipulate that cell therapeutics are processed in serum-free conditions (Ratcliffe *et al.*, 2011). Current regulatory guidelines allow the use of serum from specific sources to be prepared under strict manufacturing guidelines (EAEMP, 2004). Traceable animals from specific geographic locations are used to reduce the risk of TSE transmission to humans. Ideally, however, the cell therapy industry should move away from traditional FCS supplementation and move towards animal product-free culture methods using well-defined medium. Expansion methods using serum-free cell culture have been developed to eliminate the requirement for FCS in culture medium (Muller *et al.*, 2006).

Product safety also extends to safety of the transplanted cells or engineered tissues themselves. Risk of transplanting cells that have an unstable or altered karyotype has to be considered (Rayment and Williams, 2010) and is a particular problem with pluripotent cells explored as an allogeneic therapy where there is a risk of transplanting a differentiated target cell population that has acquired karyotypic abnormalities (Maitra *et al.*, 2005) or is contaminated with undifferentiated precursors (Halme and Kessler, 2006).

© Woodhead Publishing Limited, 2013

154   Standardisation in cell and tissue engineering

This is one fundamental reason why regulatory approval to initiate clinical trials using allogeneic, pluripotent lines has taken much longer than for both autologous and allogeneic somatic cells. However, even isolation and expansion of primary cells such as MSCs in an artificial environment provided by culture vessels *in vitro* can put selective pressure on cells that have limited replicative lifespans and that normally have low turnover rates *in vivo*, leading to genetic abnormalities (Lepperdinger *et al.*, 2008b). In particular cells from aged donors or patients whose tissues have been exposed to high levels of oxidative stress *in vivo* are prone to undergo replicative senescence or experience selective pressure in culture more quickly than cells from young patients (Lepperdinger *et al.*, 2008a) so producing an adequate quantity of cells that are genetically stable can be a challenge. It is therefore considered very important to assess genetic and epigenetic stability in established cultures of cells that will be used for therapy (Dominici *et al.*, 2006).

The FDA gives safety considerations the highest priority, so any step taken to minimise the risk of contamination is looked upon favourably. Such actions are mainly to do with process development, and include converting a closed system or eliminating animal serum, which can often carry mycoplasma, from culture media.

Purity

Purity has always been a major factor in the development of therapeutics, whether small molecules or biologics. To a large extent it goes hand in hand with safety, because having contaminants in the final product is undesirable due to the potential impact on the patient. The main difference with cell-based therapeutics rather than small molecules or biologics is that most purification steps have some sort of impact on cell phenotype, as the cells are living and hence dynamically responsive to external stimuli, and so for many candidate therapies, such a step is not desirable. Even a simple volume reduction step that involves centrifugation may expose cells to process forces that could potentially alter the product. As a result, adopting the mantra 'the product is the process' and ensuring that narrow operating parameters that have been carefully defined to deliver a high quality product are adhered to is crucial. In addition, it is important to ensure that the contents of final packaged product are free from unwanted materials used throughout the processing. In general, QC measures to ensure high purity of cell therapy products are adopted to achieve the following:

- Demonstrating that the product is free of endotoxin/pyrogen.
- Determining the level of serum albumin present (for cells processed in the presence of animal serum) as less than 1 ppm in the final formulation, as specified by The Code of Federal Regulations (FDA, 2002).

© Woodhead Publishing Limited, 2013

Quality control in cell and tissue engineering   155

- Evaluating the presence of residual harvest agents, such as trypsin.
- Proving that agents used throughout the process such as antibodies, growth factors or cytokines, are absent from the final product.

As previously discussed, it is important to understand the identity of all cell populations in the final product; with respect to purity, it is important to quantify (with justifiable tolerances) the percentage of each cell type present.

Referring once again to the mantra 'the product is the process' (Mason and Hoare, 2007), by adjusting processing parameters to levels that yield optimal purity of the target cell product, a key QC parameter is ensuring that these optimal processing parameters are adhered to. Optimising multi-factorial culture conditions can be achieved using microbioreactor arrays to identify appropriate culture conditions (Titmarsh and Cooper-White, 2009). When optimal operating conditions for the process have been defined, it is important to work within those parameters to ensure consistency in the quality of the resulting product. Automated, scalable cell culture platforms enable key processing variables to be controlled and optimal conditions maintained (Mason and Hoare, 2006). From a QC perspective, automation has the benefit of applying uniformity in processing across multiple batches or successive passages of a single batch, ensuring that a defined process is reproducibly and precisely applied over and over again, without any of the variability that often occurs during manual processing (Kino-Oka *et al.*, 2005).

The application of automation has already been explored for rapid and reliable expansion of human mesenchymal stem cells (Thomas *et al.*, 2007) and embryonic stem cells (Narkilahti *et al.*, 2007; Terstegge *et al.*, 2007; Thomas *et al.*, 2009). Direct comparison between manual and automated methods for cell expansion indicated that the automated process was superior, with manual processing suffering from high process variability (Liu *et al.*, 2010). Progression to dynamic culture systems has also proven very successful as continual perfusion of medium ensures a homogeneous environment in which factors such as oxygen, pH and nutrient/metabolite levels are carefully controlled (Ben-Amotz *et al.*, 2007). Recently, not only has perfusion *per se* been shown to enhance embryonic stem cell expansion, but the optimal perfusion rate for embryonic stem cells cultured in a perfusion microbioreactor has been determined (Hudson *et al.*, 2011). The optimal perfusion rate has to account for both appropriate duration of exposure to stimulatory media components and timely removal of inhibitory metabolites (Ben-Amotz *et al.*, 2007; Hudson *et al.*, 2011).

Certainly in the case of autologous cell therapy and to a degree between different lots of allogeneic cells, using automation to apply optimal process conditions still does not guarantee the quality of the final product due to

© Woodhead Publishing Limited, 2013

156   Standardisation in cell and tissue engineering

potential differences in the starting material, such as cell yield from biopsy, growth rates between specimens etc. (Lim *et al.*, 2007; Placzek *et al.*, 2009). However, careful bioprocess control, which can typically be achieved using automated processing, can ensure that the numbers of variables in the whole bioprocess are kept to a minimum.

Identity

Identity assays are used to confirm that the composition of a final cell therapy product is as intended, thus verifying that the cell populations have not changed unexpectedly during the manufacturing process. Such assays are crucial to prove that the product remains substantially similar after any process changes, including scaling up to larger systems. Identity testing may include an antibody-targeting method such as flow cytometry or fluorescence-activated cell sorting that specifically identifies a particular cell subset based on antibody binding to a specific defined cell surface protein and then 'sorts' cells by selectively isolating individual cells that express that surface marker (McIntyre *et al.*, 2010). Cell sorting is the most accurate method to date for identifying and purifying cell products on the basis of specific cell surface protein expression.

When flow cytometry is used as a terminal procedure for identity purposes only (not purification of a target cell product), any cells in the population that have not been labelled will generally need to be identified and justified, in order for regulatory approval to be granted. Additional methods of characterising cell identity include gene expression profiling and epigenetic profiling. The latter is particularly advantageous over simple gene expression analysis because gene expression can vary from cell to cell and even though specific gene expression signatures can be correlated to a population-wide phenotype, it does not necessarily identify a product or contaminating cell types that express the same genes. Epigenetic profiling works on the basis that different cell types have different epigenetic profiles, such as methylation patterns, and identifying the methylation pattern unique to the target cell product will provide a measure of identity and inform if contaminating cell types are present (Baron *et al.*, 2006; Rapko *et al.*, 2007)

In some cases a more qualitative approach is acceptable. For example Advanced Cell Technology's therapy to treat dry age-related macular degeneration consists of retinal pigment epithelium cells, which as one of only two cell types in the body that is pigmented (along with melanocytes) is easily distinguished from other cell types manufactured in the same facility.

Potency

Potency is notoriously the most difficult cell feature to characterise, and groups such as the Alliance for Regenerative Medicine (ARM) are actively

© Woodhead Publishing Limited, 2013

Quality control in cell and tissue engineering   157

working with regulators to try and develop reasonable methods for achieving this measure (Stroncek *et al.*, 2007). A potency assay measures the biological function of a product. Ideally, the assay should assess the final product as it is administered, and while it is preferable for testing results to correlate with functions related to clinical outcomes, this is not always possible.

Potency assays should provide a measure of biological activity, according to FDA guidelines (FDA, 2008b). Typically, quantitative assays such as enzyme-linked immunosorbant assays may be suitable for identifying levels of specific bioactive cytokines secreted from cells into the growth medium (Carmen *et al.*, 2012). Therefore if the mechanism of action of the cell therapy is established, it is possible to quantitate the potency of that. However, it may be difficult to determine potency where the mode of action remains unknown. One example of this is cardiac cell therapy, in which the proposed action mechanisms are very poorly understood (Mozid *et al.*, 2011). Conversely for tissue-engineered skin equivalents, potency might be measured simply as the ability to restore the structural barrier function of the tissue, protecting the body from the external environment, as in the case of the skin equivalent Apligraf (Trent and Kirsner, 1998; Falanga and Sabolinski, 1999).

Assays can evaluate the product quantitatively (such as enzyme-linked immunosorbant assays to measure production and secretion of specific bioactive cytokines) or qualitatively (such as an analysis of cell morphology). However due to the dynamic nature of cells, it is likely that a stability matrix of several different assays will be used to ensure that all of the critical quality attributes of the cell therapy are met (Carmen *et al.*, 2012). An exemplar stability matrix reproduced from Carmen *et al.* (2012) is presented in Table 8.1.

An additional consideration for determining potency is that one cell therapy product may be applicable to more than one clinical indication. Even though the cells have the same identity, it is different features of this identity that can be exploited for different illnesses. For this reason, cells to treat one indication may require a different concentration or format than for the same cell to treat another indication, and thus the required potency assays must also adjust accordingly. For example, Osiris Therapeutics' lead product Prochymal is implicated for several indications including Crohn's disease and acute myocardial infarction. Thus, a potency assay assessing the ability to regenerate crypt cells in the former would be inappropriate in the latter, where the aim of the product is to minimise cell death in the heart.

8.3 Commercial quality control/quality assurance in large-scale manufacture

During processing of cells for therapy, QC and QA measures have to be considered throughout the 'whole bioprocess' to ensure consistent quality

Table 8.1 Example stability matrix defining the combination of different quality tests that might be used to define release criteria for a cell therapy product

Category	Test	Specification	Lot release							Optional	
			T = 0	3 m	6 m	9 m	12 m	18 m	24 m	36 m	48 m
Viability	Cell counting Viable cell	>70%	X	X	X	X	X	X	X	X	X
Viability	Recovery	??	X				X		X		X
Safety	Sterility	Negative	X				X		X		X
Safety	Mycoplasma	Negative	X				X		X		X
Purity	Endotoxin	Negative	X				X		X		X
Identity	FACS marker #1	>95%	X	X	X	X	X	X	X	X	X
Identity	FACS marker #2	>95%	X	X	X	X	X	X	X	X	X
Identity/purity	FACS marker #3	<5%	X	X	X	X	X	X	X	X	X
Potency	Cytokine #1 ELISA	50–100 pg/cell	X	X	X	X	X	X	X	X	X
Potency	Cytokine #2 ELISA	25–75 pg/cell	X	X	X	X	X	X	X	X	X
Potency	Elicited Cytokine #3 ELISA	200–300 pg/cell	X	X	X	X	X	X	X	X	X

Source: Reproduced from Carmen et al. (2012). This table is produced with the kind permission of Jessica Carmen PhD and John Rowley PhD, both of Lonza Inc., Walkersville, MA, USA, and E. Manzotti, Future Medicine, London, UK.
ELISA - enzyme-linked immunosorbent assay; FACS - fluorescent activated cell sorting.

Quality control in cell and tissue engineering

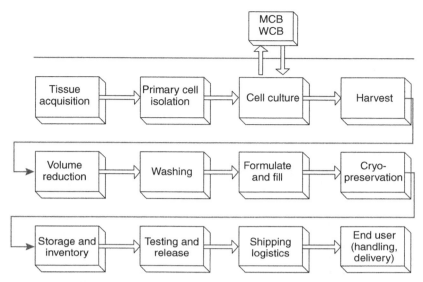

8.2 An emerging cell therapy platform process (Brandenberger *et al.*, 2011). Reprinted with permission from *BioProcess International*, March 2011, pages 30–37. MCB, master cell bank; WCB, working cell bank.

of the end product. Each stage of the process is analysed using key performance indicators, whose outcomes help to validate the QA element of processing. A generic flow sheet for cell therapy production, as outlined by Brandenberger and colleagues (2011) is presented in Fig. 8.2. It highlights the individual unit operations required to produce a cell therapy product. The QC and QA considerations that are necessary for each stage in autologous cell therapy production are presented in Table 8.2.

Whilst there are some fundamental differences between autologous and allogeneic (e.g., embryonic) cells, in terms of acquisition and potentially method of processing, the basic process outline and the QC/QA requirements are similar. For example, the need to expand cells and have a resultant population that is the same as the start material (prior to differentiation) and then, if needed, the capacity to successfully differentiate those cells into specific targets means that similar parameters have to be measured. The main consideration for cell culture is to ensure reproducibility within the process. Ensuring that each batch is processed under the same conditions for the same (optimised) number of passages/population doublings is vital for producing a consistent cell therapy. In this way, online monitoring and automation may become crucial for cell therapies in the same way that it has for biologics. It is thus expected that there will be a transition of cell culture systems from the 'wait and see' nature of cell factories to

160 Standardisation in cell and tissue engineering

Table 8.2 Key QC and QA considerations in the production of typical autologous cell-based therapies

Process step	QC measures	QA measures
Tissue acquisition	• Donor screening for transmittable diseases	• Biopsy procedure and surgical competence • Accurate documentation of relevant patient information and biopsy status
Primary cell isolation	• Cell yield	• Good laboratory practice
Cell culture	• Cell yield • Cell growth rates • Cell morphology • Metabolite analysis from culture medium	• Highly trained operators • Utilisation of automation technology for monitoring and control • Segregation of cultures to prevent cross-contamination
Harvest	• Cell yield • Cell viability tests • Intracellular protein analysis from culture medium	• Adherence to highly defined, reproducible and validated protocols formulated based on a clear understanding of the impact of enzyme exposure time
Volume reduction	• Cell yield • Cell viability tests	• Adherence to validated protocols that have been developed to account for shear stress during centrifugation or filtration
Washing	• Absence of contaminating residual culture medium reagents or trypsin	• Adherence to defined, validated protocols
Formulate and fill	• Cell concentration • DMSO concentration (if used)	
Cryopreservation	• Cell yield and viability upon thawing	• Freezing time and temperature determined and validated • Appropriate thawing time and rate of thawing determined and then adhered to
Storage and inventory		• Consistent labelling and tagging
Testing and release	• Cell viability • Cell identity (e.g., flow cytometry) • Cell potency (e.g., ELISA for secreted factors) • Sterility testing	
Shipping logistics	• Cell viability and potency in extremes of temperature, time in transit, etc.	• Regulation of external environmental conditions (e.g., cold shipper boxes for frozen products)
End-user and administration		• Clinical competency and defined protocols

DMSO, dimethyl sulfoxide; ELISA, enzyme-linked immunoabsorbent assay.

© Woodhead Publishing Limited, 2013

Quality control in cell and tissue engineering 161

the carefully controlled environments of plated or suspension bioreactors (Brandenberger *et al.*, 2011). One principle difference between autologous and allogeneic cell therapy is that in allogeneic cell culture, one room or manufacturing space will be used to make one product. Autologous products require a much greater degree of segregation. However it is impractical to have one room per patient and so precautions must be taken to ensure that there is no cross-contamination between patients and advances in automation may provide suitable solutions here (Mason and Hoare, 2007; Ratcliffe *et al.*, 2011). Furthermore, in autologous cells there may be instances where cells taken from a patient fail to grow or are contaminated. This generally accounts for less than 1% of samples taken (Mayhew *et al.*, 1998; Mason and Hoare, 2007) but it is important that the process is sufficiently defined that failure of cell growth does not bring into question process stability.

The key difference between autologous and allogeneic cells arises in terms of acquisition of cells. The donor cells that will form an allogeneic cell bank will have been thoroughly screened for many different characteristics. Generally at this stage the QC steps depend both on the desired cell type and the method of acquisition.

In autologous tissue acquisition, the sample may be taken at a hospital/doctor's surgery (e.g., apheresis, bone marrow aspirate) or at a specialised treatment centre by specially trained staff. At this point, although it is desirable to maintain the highest level of sterility possible, the main QC concern is to ensure that the sample is properly recorded. Each specimen is labelled with a unique ID code that links it electronically to the corresponding patient. This code is used to track the sample throughout the entire process, using technology such as the MODA paperless QC system utilised by Lonza Biologics (Lonza, 2011). The thorough documentation of samples throughout the process is hugely important in quality management; this is also true for single donor-to-patient therapies (such as donor-matched cord blood therapies).

For allogeneic products, where the same single cell line is used to treat thousands of patients, there is an enormous need to ensure product consistency between batches. A master cell bank (MCB) of homogeneous cell aliquots is used to capture a single defined phenotype, enabling the repeated formation of functionally identical cells. This MCB is maintained in a frozen state, is incredibly well characterised and limits the number of passages that cells undergo. This is important, as senescence in cells is associated with increased karyotypic instability.

The master cell bank gives rise to all eventual therapies via smaller working cell banks (WCBs), which enable the controlled expansion of cells. It is imperative that the master cell bank is scrupulously characterised, as any imperfections are amplified as the cells undergo further population doublings. For this reasons, the source of the cells, whether it be an adult donor or embryo,

© Woodhead Publishing Limited, 2013

162 Standardisation in cell and tissue engineering

is rigorously assessed before an MCB is created. The QC measures taken include cytogenetic studies, DNA profiling, isoenzyme analysis, morphological assessment and analysis of surface markers. WCBs must also undergo QC procedures, although to a lesser extent, with a principle focus on identity and safety.

8.4 Conclusion

The importance of carefully considered QC/QA strategies is fundamental for cell therapy production, primarily because cells are living and can undergo responsive changes to stimuli they receive during bioprocessing. QC measures can be used to ensure that the cellular material is of the desired quality but often in cases where cell material is scarce, such as in the case of autologous cell therapy, manufacturers become more reliant upon adopted QA strategies. QA strategies facilitate robust demarcation of the process steps, to ultimately ensure that the cells have been treated identically and should therefore have similar cellular phenotypes.

8.5 References

Baron, U., Turbachova, I., Hellwag, A., Eckhardt, F., Berlin, K., Hoffmuller, U., Gardina, P. and Olek, S. 2006. DNA methylation analysis as a tool for cell typing. *Epigenetics: Official Journal of the DNA Methylation Society*, **1**, 55–60.

Ben-Amotz, R., Lanz, O. I., Miller, J. M., Filipowicz, D. E. and King, M. D. 2007. The use of vacuum-assisted closure therapy for the treatment of distal extremity wounds in 15 dogs. *Veterinary Surgery: VS*, **36**, 684–690.

Brandenberger, R., Burger, S., Campbell, A., Fong, T., Lapinskas, E. and Rowley, J. 2011. Cell therapy bioprocessing: integrating process and product development for the next generation of biotherapeutics. *BioProcess International*, **9**, 30–37.

Carmen, J., Burger, S. R., Mccaman, M. and Rowley, J. A. 2012. Developing assays to address identity, potency, purity and safety: cell characterization in cell therapy process development. *Regenerative Medicine*, **7**, 85–100.

Dominici, M., Le Blanc, K., Mueller, I., Slaper-Cortenbach, I., Marini, F., Krause, D., Deans, R., Keating, A., Prockop, D. and Horwitz, E. 2006. Minimal criteria for defining multipotent mesenchymal stromal cells. The International Society for Cellular Therapy position statement. *Cytotherapy*, **8**, 315–317.

EAEMP 2004. Note for guidance on minimising the risk of transmitting animal spongiform encephalopathy agents via human and veterinary medicinal products. London: European Agency for the Evaluation of Medicinal Products.

Falanga, V. and Sabolinski, M. 1999. A bilayered living skin construct (APLIGRAF) accelerates complete closure of hard-to-heal venous ulcers. *Wound repair and regeneration: official publication of the Wound Healing Society [and] the European Tissue Repair Society*, **7**, 201–207.

Food and Drug Administration. 2002. *General Biological Products Standards*. Rockville: United States Food and Drug Administration.

© Woodhead Publishing Limited, 2013

Food and Drug Administration. 2008a. *Draft Guidance for Industry: Validation of Growth-based Rapid Microbiological Methods for Sterility Testing of Cellular and Gene Therapy Products.* Rockville: United States Food and Drug Administration.

Food and Drug Administration. 2008b. *Guidance for FDA Reviewers and Sponsors: Content and Review of Chemistry, Manufacturing and Control (CMC) Information for Human Somatic Cell Therapy Investigational New Drug Applications (INDs).* Rockville: United States Food and Drug Administration.

Food and Drug Administration. 2011. *General Biological Products Standards.* Rockville: United States Food and Drug Administration.

Goldring, C. E., Duffy, P. A., Benvenisty, N., Andrews, P. W., Ben-David, U., Eakins, R., French, N., Hanley, N. A., Kelly, L., Kitteringham, N. R., Kurth, J., Ladenheim, D., Laverty, H., Mcblane, J., Narayanan, G., Patel, S., Reinhardt, J., Rossi, A., Sharpe, M. and Park, B. K. 2011. Assessing the safety of stem cell therapeutics. *Cell Stem Cell*, **8**, 618–628.

Halme, D. G. and Kessler, D. A. 2006. FDA regulation of stem-cell-based therapies. *The New England Journal of Medicine*, **355**, 1730–1735.

Hashii, N., Kawasaki, N., Nakajima, Y., Toyoda, M., Katagiri, Y., Itoh, S., Harazono, A., Umezawa, A. and Yamaguchi, T. 2007. Study on the quality control of cell therapy products. Determination of N-glycolylneuraminic acid incorporated into human cells by nano-flow liquid chromatography/Fourier transformation ion cyclotron mass spectrometry. *Journal of Chromatography A*, **1160**, 263–269.

Higano, C. S., Small, E. J., Schellhammer, P., Yasothan, U., Gubernick, S., Kirkpatrick, P. and Kantoff, P. W. 2010. Sipuleucel-T. *Nature Reviews. Drug Discovery*, **9**, 513–514.

Hudson, J., Titmarsh, D., Hidalgo, A., Wolvetang, E. and Cooper-White, J. 2011. Primitive cardiac cells from human embryonic stem cells. *Stem Cells and Development*, **21**, 1513–1523.

Kielpinski, G., Prinzi, S., Duguid, J. and Du Moulin, G. 2005. Roadmap to approval: use of an automated sterility test method as a lot release test for Carticel, autologous cultured chondrocytes. *Cytotherapy*, **7**, 531–541.

Kino-Oka, M., Ogawa, N., Umegaki, R. and Taya, M. 2005. Bioreactor design for successive culture of anchorage-dependent cells operated in an automated manner. *Tissue Engineering*, **11**, 535–545.

Lepperdinger, G., Brunauer, R., Gassner, R., Jamnig, A., Kloss, F. and Laschober, G. T. 2008a. Changes of the functional capacity of mesenchymal stem cells due to aging or age-associated disease – implications for clinical applications and donor recruitment. *Transfusion Medicine and Hemotherapy: Offizielles Organ der Deutschen Gesellschaft fur Transfusionsmedizin und Immunhamatologie*, **35**, 299–305.

Lepperdinger, G., Brunauer, R., Jamnig, A., Laschober, G. and Kassem, M. 2008b. Controversial issue: is it safe to employ mesenchymal stem cells in cell-based therapies? *Experimental Gerontology*, **43**, 1018–1023.

Lim, M., Ye, H., Panoskaltsis, N., Drakakis, E. M., Yue, X., Cass, A. E., Radomska, A. and Mantalaris, A. 2007. Intelligent bioprocessing for haemotopoietic cell cultures using monitoring and design of experiments. *Biotechnology Advances*, **25**, 353–368.

164   Standardisation in cell and tissue engineering

Liu, Y., Hourd, P., Chandra, A. and Williams, D. J. 2010. Human cell culture process capability: a comparison of manual and automated production. *Journal of Tissue Engineering and Regenerative Medicine*, **4**, 45–54.

Lonz, A. 2011. MODA: Paperless QC Micro (Online). Available: http://www.lonzabio.com/2461.html (Accessed 1 December 2011).

Maitra, A., Arking, D. E., Shivapurkar, N., Ikeda, M., Stastny, V., Kassauei, K., Sui, G., Cutler, D. J., Liu, Y., Brimble, S. N., Noaksson, K., Hyllner, J., Schulz, T. C., Zeng, X., Freed, W. J., Crook, J., Abraham, S., Colman, A., Sartipy, P., Matsui, S., Carpenter, M., Gazdar, A. F., Rao, M. and Chakravarti, A. 2005. Genomic alterations in cultured human embryonic stem cells. *Nature Genetics*, **37**, 1099–1103.

Mason, C. and Dunnill, P. 2007. Translational regenerative medicine research: essential to discovery and outcome. *Regenerative Medicine*, **2**, 227–229.

Mason, C. and Hoare, M. 2006. Regenerative medicine bioprocessing: the need to learn from the experience of other fields. *Regenerative Medicine*, **1**, 615–623.

Mason, C. and Hoare, M. 2007. Regenerative medicine bioprocessing: building a conceptual framework based on early studies. *Tissue Engineering*, **13**, 301–311.

Mayhew, T. A., Williams, G. R., Senica, M. A., Kuniholm, G. and Du Moulin, G. C. 1998. Validation of a quality assurance program for autologous cultured chondrocyte implantation. *Tissue Engineering*, **4**, 325–334.

Mcintyre, C., Flyg, B. and Fong, T. 2010. Fluorescence-activated cell sorting for cGMP processing of therapeutic cells. *BioProcess International*, **8**, 44–53.

Mozid, A. M., Arnous, S., Sammut, E. C. and Mathur, A. 2011. Stem cell therapy for heart diseases. *British Medical Bulletin*, **98**, 143–159.

Muller, I., Kordowich, S., Holzwarth, C., Spano, C., Isensee, G., Staiber, A., Viebahn, S., Gieseke, F., Langer, H., Gawaz, M. P., Horwitz, E. M., Conte, P., Handgretinger, R. and Dominici, M. 2006. Animal serum-free culture conditions for isolation and expansion of multipotent mesenchymal stromal cells from human BM. *Cytotherapy*, **8**, 437–444.

Narkilahti, S., Rajala, K., Pihlajamaki, H., Suuronen, R., Hovatta, O. and Skottman, H. 2007. Monitoring and analysis of dynamic growth of human embryonic stem cells: comparison of automated instrumentation and conventional culturing methods. *Biomedical Engineering Online*, **6**, 11.

Placzek, M. R., Chung, I. M., Macedo, H. M., Ismail, S., Mortera Blanco, T., Lim, M., Cha, J. M., Fauzi, I., Kang, Y., Yeo, D. C., Ma, C. Y., Polak, J. M., Panoskaltsis, N. and Mantalaris, A. 2009. Stem cell bioprocessing: fundamentals and principles. *Journal of the Royal Society, Interface/The Royal Society*, **6**, 209–232.

Rapko, S., Baron, U., Hoffmuller, U., Model, F., Wolfe, L. and Olek, S. 2007. DNA methylation analysis as novel tool for quality control in regenerative medicine. *Tissue Engineering*, **13**, 2271–2280.

Ratcliffe, E., Thomas, R. J. and Williams, D. J. 2011. Current understanding and challenges in bioprocessing of stem cell-based therapies for regenerative medicine. *British Medical Bulletin*, **100**, 137–155.

Rayment, E. A. and Williams, D. J. 2010. Concise review: mind the gap: challenges in characterizing and quantifying cell- and tissue-based therapies for clinical translation. *Stem cells*, **28**, 996–1004.

Rohde, E., Schallmoser, K., Bartmann, C., Reinisch, A. and Strunk, D. 2008. GMP-compliant propagation of human multipotent mesenchymal stromal cells. In: Gad, S. (ed.) *Pharmaceutical Manufacturing Handbook: Regulations and Quality*. Hoboken, New Jersey: John Wiley & Sons.

Quality control in cell and tissue engineering 165

Sampath, R., Blyn, L. B. and Ecker, D. J. 2010. Rapid molecular assays for microbial contaminant monitoring in the bioprocess industry. *PDA Journal of Pharmaceutical Science and Technology/PDA*, **64**, 458–464.

Stacey, A. and Stacey, G. 2000. Routine quality control testing of cell cultures: detection of Mycoplasma. *Methods in Molecular Medicine*, **24**, 27–40.

Stacey, G. and Auerbach, J. 2007. Quality control procedures for stem cell lines. In: Freshney, R., Stacey, G. and Auerbach, J. (eds) *Culture of Human Stem Cells*. Hoboken, New Jersey: John Wiley & Sons, Inc.

Stroncek, D. F., Jin, P., Wang, E. and Jett, B. 2007. Potency analysis of cellular therapies: the emerging role of molecular assays. *Journal of Translational Medicine*, **5**, 24.

Terstegge, S., Laufenberg, I., Pochert, J., Schenk, S., Itskovitz-Eldor, J., Endl, E. and Brustle, O. 2007. Automated maintenance of embryonic stem cell cultures. *Biotechnology and Bioengineering*, **96**, 195–201.

Thomas, R. J., Anderson, D., Chandra, A., Smith, N. M., Young, L. E., Williams, D. and Denning, C. 2009. Automated, scalable culture of human embryonic stem cells in feeder-free conditions. *Biotechnology and Bioengineering*, **102**, 1636–1644.

Thomas, R. J., Chandra, A., Liu, Y., Hourd, P. C., Conway, P. P. and Williams, D. J. 2007. Manufacture of a human mesenchymal stem cell population using an automated cell culture platform. *Cytotechnology*, **55**, 31–39.

Titmarsh, D. and Cooper-White, J. 2009. Microbioreactor array for full-factorial analysis of provision of multiple soluble factors in cellular microenvironments. *Biotechnology and Bioengineering*, **104**, 1240–1244.

Trent, J. F. and Kirsner, R. S. 1998. Tissue engineered skin: Apligraf, a bi-layered living skin equivalent. *International Journal of Clinical Practice*, **52**, 408–413.

© Woodhead Publishing Limited, 2013

9
Standardised chemical analysis and testing of biomaterials

W. CHRZANOWSKI and F. DEHGHANI,
The University of Sydney, Australia

DOI: 10.1533/9780857098726.2.166

Abstract: This chapter presents the most recent advances in analysis of modern biomaterials according to the standards provided by several organisations. In the chapter the reader will find information on how to evaluate physico-chemical and biological activity of materials for biomedical applications. The main focus is on surface and bulk chemical analysis, physical and structural examinations and biocompatibility assessment that are key factors for approval of biomaterials that are designed for applications in the body.

Key words: chemical analysis, structural analysis, ASTM and ISO standards.

9.1 Introduction: why we need standard methods for testing biomaterials

Biomedical engineering has had a significant impact on the improvement and progress of modern medicine. Advances in technology for the design, manufacture and testing of implants and medical devices have enhanced both the quality of life and life expectancy of patients. Testing of these materials plays a critical role in reducing the risk of failure in animal studies, clinical trial and the final products. The testing of biomaterials involves assessing the safety, toxicity, efficacy and purity of a product. As biomaterials are designed to interact with the human body, it is critical that moral and ethical concerns are addressed prior to *in vivo* tests and clinical trials, and that legislation governing the development of biomedical products is closely followed.

Biomaterials are still an emerging field, necessitating a huge number of animal studies and clinical trials. However, advanced and standard characterisation techniques for physico-chemical analysis and *in vitro* studies can significantly reduce the number of animal and human trials needed, increase the success rate of these experimentations, and therefore decrease the cost of product development. A number of organisations have made available worldwide regulatory frameworks and guidelines for the analysis of biomaterials, including the International Organization for Standardization (ISO), the American Society for Testing and Materials (ASTM), United States Pharmacopeia (USP) and British Pharmacopoeia (BP). Companies are restricted to the use of these standard methods and invest millions of dollars in testing for quality control to prevent the risk of product failure.

Whilst there is a lack of formal ethical training for biomedical engineers, there is a great need for the scientists who work in this field to produce professional, trustworthy results, conducting non-biased tests to thoroughly investigate all possible concerns about the future safety of any products in development.

9.2 Standardised chemical analysis: when and why we assess chemistries

Chemical composition has a significant impact on the performance of biomedical devices. Chemical cues, next to topographical and mechanical, are the most important factors in the regulation of cellular responses. Different chemistries can be used to regulate and identify the safety, bioactivity and antimicrobial properties of a material. It is important to note that chemistry refers to both organic (e.g., protein or synthetic polymer) and inorganic (elemental composition) phases that constitute a material. In addition to the individual molecules and elements, the molecular structure and type of compound also play a critical role in the properties of a biomaterial. Alloys that contain a significant amount of titanium, for example, exhibit superior biological properties due to a layer of active titanium oxide which forms on their surface. This oxide layer (TiO_2) promotes positive responses in the body environment that may not be highlighted in studies of the single element. Hence probing the individual element but not the compound could be misleading, and this issue has to be taken into consideration when designing studies.

A concern for the safety and efficacy of a product develops when the level of impurities is beyond the acceptable level recommended by the US Food and Drug Administration (FDA) for human application. The source of impurities for natural and synthetic biomaterials can be biological, for materials such as proteins (e.g., silk, collagen and elastin) and polysaccharides (e.g., cellulose, alginate, hyaluronan, and chitin), but can also include degradation products, residual solvents, unreacted crosslinking agents, monomers and residues transferred from such processing stages as sterilisation. Bulk properties of materials are typically different from their surface properties, and special analytical techniques are required for the characterisation of each part. The surface plays the most important role in interactions with the biological environment. Analysis of surface chemistry can therefore assist

168 Standardisation in cell and tissue engineering

in determining mechanisms that regulate cellular/bacterial responses. The corrosion and degradation of metal, glasses and polymers usually focuses on the material as a bulk structure, yet surface modification such as coating and painting can inhibit corrosion. However, it is also critical to analyse the bulk chemistry of a biomaterial, as this has a significant impact on a material's mechanical properties, degradation, thermal and electrical characteristics. In many materials (e.g., glasses) degradation occurs in the bulk and affects the internal chemical structure. However, in general the type of material and its application can best be investigated by analysis of both the surface and bulk of the material. This approach is applicable to a broad group of biomaterials.

9.2.1 Analysis of surface and bulk chemistry

It is known that many materials must be regulated in terms of chemical composition if they are to be safely used in the body. However, in testing biomaterials it is important to measure both bulk and surface chemistries. Generally, elemental analysis is performed in addition to tests for characterisation of the compound and structure. The bulk chemistry of a biomaterial is always measured to confirm safety and purity, whilst surface analysis reveals the interplay between the material (i.e., the surface) and the biological milieu. There is still no standard method available for determining the surface properties *in vitro*, due to the difficulty in regulating the composition of each material after surface modification has taken place. The standards therefore regulate the responses of cells and bacteria to such surfaces, which include cytotoxicity, sensitisation, irritation, systemic toxicity, subchronic toxicity, genotoxicity, implantation and haemocompatibility.

Chemical analyses are highly relevant for virtually all materials used in medicine, with both bulk and surface examinations providing essential information. Compounds selected for the modification of surfaces need to comply with standards for safety, and should be chosen to prevent any side-effects or complications that may arise from the application of a product and any subsequent degradation residues.

9.3 Chemical properties

The standard analytical methods recommended for chemical analysis of biomaterials are listed in Table 9.1. In addition, the USP number that describes the details of each method is included in the table. These methods can be used for the detection of impurities and the identification of desirable compounds.

Well-established methods for detecting impurities and confirming the purity of a product include gas chromatography (GC); high-performance liquid

Standardised chemical analysis and testing of biomaterials 169

Table 9.1 USP chemical tests

USP test number	Test description	Analytical techniques
<197>	Spectrophotometeric identification	UV *vis*, FTIR
<231>	Heavy metals	Inductively coupled plasma (ICP), atomic adsorption spectroscopy
<381>	Elastomeric closure for injection – physicochemical test	
<731>	Loss in drying (water content)	Thermogravimetry (TGA), differential scanning calorimetry (DSC)
<736>	Mass spectroscopy – purity or elemental analysis	HPLC/MS
<761>	Nuclear magnetic resonance – purity or component analysis (e.g. copolymers)	NMR
<851>	Spectrophotometry and light scattering (molar mass information)	
<891>	Thermal analysis (purity)	TGA, DSC
<911>	Viscosity (molar mass)	

Source: Adapted from ASTM F2150, 2007.

chromatography and mass spectrometry (HPLC/MS); UV and Fourier transform infrared (FTIR) spectroscopic techniques; nuclear magnetic resonance (NMR); gel electrophoresis (SDS-PAGE) for protein analysis; and gel permeation chromatography (GPC). Mass spectrometry is an accurate technique for measuring the mass of compounds, whilst thermal analysis (differential scanning calorimeter and thermal gravimetric analysis) can be used for detecting the degree of purity of a material. Users need to take into account the detection limit of each method for a quantitative analysis, to ensure that the minimum value is well below the acceptable level. The detection limit for identifying any impurities is dependent on the analytical method selected. It is important to report the detection limit of any method used for the quantitative analysis of impurities in an appropriate unit, such as ppm (μg/g, mg/kg) or ppb (ng/g, μg/kg).

Any chemical or contaminant residues that are not part of the intended design are considered to be impurities. The source of impurities may be the raw material itself, solvents or chemicals used during the processing and manufacture of a product. Acceptable levels are a function of the type of impurity and the intended use of the material. As previously noted, impurities are observed in both natural-based materials, such as proteins (elastin and collagen) and polysaccharides (cellulose, alginate, chitosan, hyaluronan) as well as in synthetic materials (solvent, monomer, crosslinking agent, metal residues, endotoxins, agents used for sterilisation).

Impurities separated in gel electrophoresis can be detected by Coomassie Blue (as a general protein stain) or silver (as a general protein and carbohydrate stain). Further characterisation can be provided by immunoblot

© Woodhead Publishing Limited, 2013

170   Standardisation in cell and tissue engineering

analysis and/or protein sequencing to identify specific impurities that may produce critical biological activities (for example, elastin immuno-genicity, cytokines and growth factors). Other sensitive methods such as enzyme-linked immuno sorbent assay (ELISA) can be used for detection of these impurities. GC in combination with FTIR and MS can be used for qualitative and quantitative detection of volatile compounds with relatively low molecular weights (MWs).

9.3.1 Molecular weight

Gas permeation chromatography can be used to determine the MW of a poly-meric compound. If the material cannot be dissolved in a solvent suitable for GPC analysis, other techniques such as intrinsic viscosity (IV), light-scattering or membrane osmometry may be used. SDS-polyacrylamide gel electrophore-sis (SDS-PAGE) has also been used to measure protein MW and impurities. GPC and IV are suitable methods for studying linear polymers, whilst the use of light-scattering techniques is recommended for branched polymeric systems.

9.3.2 Crystallinity of biomaterial

Crystallinity can affect biomaterial characteristics, and the mechanical prop-erties of an implant or scaffold may potentially be significantly altered by crystallinity. Methods such as X-ray diffraction (XRD), differential scan-ning calorimetry (DSC) and FTIR technology can be used to determine the crystallinity of materials.

XRD measures the distance between successive atomic planes and posi-tions of atoms or ions within a crystal, allowing for determination of the crystal structure. XRD is used for analysis of solid materials and powders, and data generated can be used to fingerprint minerals whilst assessing crys-tal structure and symmetry. The relative proportion of mineral mixtures is measured from XRD analysis by comparing diffraction line intensity. This technique can be used at both high and low temperatures, making it a very useful tool for the assessment of structural changes in different materials as a function of temperature. This approach is particularly useful in shape memory materials, where changes to the crystal structure (phase transfor-mation) play a major role in programmable deformation.

In principle, substrates that contain less than 3–5 wt% crystalline phases cannot be detected using a bench-top XRD. Similarly, mixtures of phases with low symmetry are difficult to differentiate by XRD due to the larger number of diffraction peaks they produce.

The mechanical properties of many biomaterials are a function of glass transition temperatures (T_g), melting temperatures (T_m), and crystallinity.

© Woodhead Publishing Limited, 2013

It is therefore necessary to measure these properties prior to manufacturing any product, to ensure consistency in mechanical properties. Methods such as DSC can be used to measure thermal and crystalline properties of polymers or ceramics. Dynamic vapour sorption (DVS) is another quantitative method that can be used to determine the degree of crystallinity of a powder. These methods are applicable primarily for glass, ceramic, polymers and pharmaceutical products.

The density of a material may therefore be determined by a range of analytical techniques that report porosity, or draw on standard USP techniques.

9.3.3 Elemental composition

Electron probe X-ray microanalysis techniques (EDX), are the most commonly used methods for surface analysis (Table 9.3). These include energy dispersive spectroscopy (EDS), wavelength-dispersive X-ray spectroscopy (WDS) and XRD. Typically, the X-ray penetration depth is several to tens of micrometers, providing information on the elemental composition (EDS, WDS), structure (XRD) and crystallinity of a sample.

EDS and WDS can be used to assess homogeneity in terms of the composition and concentration of a coating, any hybrids used, and composite materials. In Plates V and VI (see the colour section between pages 134 and 135) electrospun PLA scaffolds employing calcium phosphate glass fillers were mapped using EDS. These images show the distribution of the elements within the structure, allowing conclusions to be reached regarding non-uniform distribution and filler particle sizes. The data in Plate V can be used to determine the composition of each element.

EDX uses the characteristic X-rays generated from a sample bombarded with electrons to identify the elemental composition. Elements can be identified from peaks corresponding to specific X-ray lines that are generated by the technique. This method can be used for quantitative analysis and comparison of a product, greatly enhancing quality control. The heights or areas under specific peaks can be used to compare a product with a reference material. EDS is considered to be the more rapid analytical method in comparison to WDS and is commonly used for quality control, as it allows a complete spectrum of energies to be acquired simultaneously. However, in the WDS spectrum the full wavelength range is scanned and acquired sequentially. Furthermore, the resolution of EDS is compromised, as the typical resolution of an ED detector is 70–130 eV, compared to peak widths in WD of 2–20 eV.

The detection limit of elemental concentrations in WDS is typically an order of magnitude lower than that acquired using EDS. This makes WDS particularly appropriate in trace elemental analysis, for which the resolution of the EDS detector may not be adequate due to overlap of adjacent peaks.

172   Standardisation in cell and tissue engineering

In practice EDS can be used for an initial survey of an unknown sample, and WDS can subsequently be employed to acquire more accurate data. In other words, EDS can be used for quantification of the elements present in relatively large amounts, while WDS allows the quantification of trace elements present at a composition below 1 wt%.

EDS and WDS are not well suited for analysis of the very thin films of organic/inorganic materials commonly used to enhance biological activity. The most common method for analysis of such surfaces is X-ray photoelectron spectroscopy (XPS) and thin film XRD (TF-XRD). XPS is a very powerful technique which provides detailed information about the composition of surfaces and materials at a depth of 4–7 nm. It is possible to produce a depth profile by sputtering the surface with argon and detecting the composition at different levels. This approach however does face the limitation that each element is sputtered at a different rate, meaning small deviations are to be expected.

The unique feature of XPS is its capacity for analysis of thin layers of protein and even DNA. This is achieved by the deconvulation of specific peaks (e.g., N, C, O) and analyses of the specific binding energies that are associated with each compound. XPS works by irradiating a sample material with X-rays, causing electrons to be ejected from the specimen. Atomic elements in the sample can be identified from the kinetic energies of these ejected photoelectrons, and their concentrations established from the photoelectron intensities.

The main features and advantage of XPS spectroscopy include:

- Allows the detection of elements from Li to U.
- Is a non-destructive technique (although may cause some damage to X-ray beam-sensitive materials).
- Can be used for chemical state analysis (some exceptions).
- Surface sensitivity is within 5–8 nm.
- Can be used for both conducting and insulating materials.
- Detection limits are in the range of 0.01–0.5 atom percent.
- Spatial resolution for surface mapping commences from 10 mm.
- Possesses depth profiling capability.

9.3.4 Fourier transform infrared (FTIR) spectroscopy

Fourier transform infrared spectroscopy is used to probe the molecular fingerprint of materials, allowing close examination of a sample's chemical make-up. FTIR spectroscopy studies the interactions of infrared radiation with a sample, which could be a solid, liquid or gas. Since chemical functional groups absorb radiation at specific frequencies by measuring these frequencies and intensities, it is possible to harness this characteristic

Standardised chemical analysis and testing of biomaterials 173

in identifying the chemistry of a sample. The concentration of a component can be calculated by measuring its absorption intensity. For a pure compound, this plot is like a molecular fingerprint because of its unique characteristics. FTIR can be used in combination with microscopy (called FTIR microscopy) to create two-dimensional maps of surface chemistries. This is of particular interest in the study of composite materials, coating and surface modifications, where it can be used to examine uniformity and distribution of specific functional groups of chemical compounds in the structure. The FTIR resolution is limited to the spot size of the laser beam, which is around 1 μm. The main advantages of the technique are that it provides information on molecular and compound composition (making it a useful tool for organic analysis) and can be used for virtually all types of material in all states (including transparent and opaque). Furthermore, it is a non-destructive technique which can be conducted at room temperature and provides both qualitative and quantitative information. The main drawbacks are the compromised resolution and large sampling depth (>100 nm), which makes assessment of nanomaterials challenging. FTIR does not provide elemental information and is not commonly used for analysis of inorganic compounds.

The most popular modes of operation are absorption or transmission. Attenuated total reflectance FTIR (ATR-FTIR) is particularly well suited for biomedical applications, providing rapid analysis of unknown samples prepared from thin films and coatings (Fig. 9.1). ATR operates by measuring the changes that occur in a totally internally reflected infrared beam when the beam comes into contact with a sample (Plate VI). An infrared beam is directed onto an optically dense crystal with a high refractive index at a certain angle. This internal reflectance creates an evanescent wave that extends beyond the surface of the crystal into the sample (which is held in contact with the crystal). This evanescent wave protrudes only a few microns (0.5–5 μm) beyond the crystal surface into the sample. In regions of the infrared spectrum where the sample absorbs energy, the evanescent wave will be attenuated or altered. The attenuated energy from each evanescent wave is passed back to the IR beam, which then exits the opposite end of the crystal and is passed to the detector in the IR spectrometer. The system then generates an infrared spectrum.

9.4 Imaging methods for measuring porosity

Tissue engineering is an alternative approach to xenografts, heterografts and the donor-sourced materials that are subject to increasing shortages. Significant advantages are offered by using a patient's own cells to produce new tissue, as this reduces the risk of failure due to rejection by the immune system. A porous scaffold is commonly used to culture harvested cells from

© Woodhead Publishing Limited, 2013

174   Standardisation in cell and tissue engineering

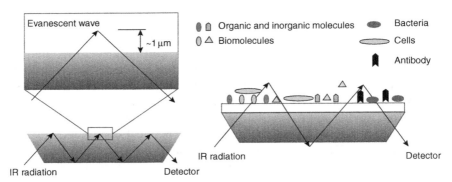

9.1 Schematic representation of the operation of ATR-FTIR.

a patient. The microstructure (porosity) of such scaffolds can significantly affect cell responses (cell migration, growth, proliferation and adhesion) and the mechanical strength of biomaterials. The establishment of a robust technique for the creation of desirable porosity and the development of reliable methods for the precise measurement of these pores are key challenges faced by biomedical engineering. Factors such as the chemical composition of a biomaterial, surface chemistry, microstructure created in scaffold, media, number of cell passages and mechanical stimulation can all affect cell behaviour. It is important to note that the definition of pore size range depends on the field of study. For tissue engineering, medical implants, and diagnostic or biological filtration applications, nanopores, micropores and macropores are within the range of 2–100 nm, 0.1–100 μm (most frequently 0.1–20 μm), and > 100 μm, respectively.

The methods developed for creating porosity in polymers include, but are not limited to, solvent casting/salt leaching, freeze-drying, gas foaming, phase separation and micropatterning. The pore structures and shape may not be uniform in the material, and pores may be either open-ended (through pore), blind-end or closed, as depicted in Fig. 9.2.

The techniques that are currently used for physical characterisation of macro/micro/nano-sized pores include microscopy (electron, optical, confocal), MicroCT scanning, magnetic resonance imaging (MRI), density measurement for pore volume, porosimetry, porometry, and NMR. An average value is generally reported for porosity and pore size (Table 9.2). Furthermore, the general abbreviations for the various methods used are highlighted later in (Table 9.3). Scanning electron microscopy (SEM) and transmission electron microscopy (TEM) are generally used to measure pore size distribution using Image J software and a statistical approach. However, for quality control and systematic study to determine the impact of pore size on cell behaviour, it is critical to use a range of different analytical techniques for pore size measurement. Different

Table 9.2 Pore size characterisation techniques

Pore size range	Selected method of analysis
Nano-pores: 0.0001–0.1 µm	Scanning tunnel microscopy, adsorption/condensation, thermoporometry/Hg intrusion, X-ray scattering, electron microscopy
Micro-pores: 0.1–1 µm	Electron microscopy, fluid flow, Hg intrusion
1–10 µm	Optical imaging, mechanical tracing, fluid flow, X-ray scattering, Hg intrusion
10–100 µm	MicroCT, thermoporometry (Hg intrusion), optical imaging, mechanical tracing
Macro-pores: >100 µm	MicroCT, thermoporometry/Hg intrusion, holography (optical imaging), mechanical tracing

9.2 Schematic of different types of pores.

porosity values may be measured by different analytical methods, due to the varied approaches used in each technique. Porosimetry, for example, is sensitive to both through and blind-end pores, whilst density can be used for measuring through pores, blind-end and closed pores too. The differences between these methods depend on the percentage of the different pore types present in a sample and the dimensions of these pores.

9.4.1 Scanning electron microscopy (SEM)

The SEM method can be used for measuring the shape, average size and distribution of pores. The permeability and tortuosity can be measured by 3D virtual images generated using this technique. However, when preparing samples for measuring the porosity of soft tissues and hydrogels using liquid nitrogen and freeze-drying, care must be taken to avoid slow cooling rates. The sample may otherwise be damaged by the thin layer of nitrogen gas produced, the formation of which results in a slow cooling rate. It is critical to freeze these samples in slush nitrogen to reduce the thickness of any insulating gas layer and enhance the cooling rates. Alternatively, cryogenic SEM may overcome this issue as it is not necessary to freeze-dry the sample using this process.

9.4.2 Transmission electron microscopy (TEM)

Transmission electron microscopy (TEM) can be used to measure the porosity of most polymer-based scaffolds. This is conducted by sectioning and mounting a sample in epoxy resin using a standard procedure employed for several examinations by TEM. For some samples, there might be an issue regarding contrast in the TEM method. In such case, samples can be stained using chemicals such as osmium tetroxides which react with carbon-carbon double bonds. TEM is not recommended for assessing the porosity of hydrogels, as the sample preparation involves dehydration for mounting the specimen in a resin. Such dehydration affects the actual pore size of a hydrogel.

9.4.3 Optical microscopy

Optical microscopy-based methods do not require any sample preparation. However, such methods are only suitable for analysis of thin samples (<0.5 mm) where adequate contrast exists between the structure and the surrounding media for the surface features to be studied. Light microscopy can be used for the characterisation of pores in a transparent sample, allowing further assessment of pore interconnectivity. However, confocal microscopy offers the advantage that it can capture images at different depths. Several laser light sources are used in modern equipment, such as laser scanning confocal microscopy (LSCM) to improve sensitivity.

9.4.4 Optical coherence tomography

Optical coherence tomography is based on interferometry rejection of out-of-plane scattering of photons. This technique possesses a high sensitivity (90 dB), and has been discussed in detail in the *Handbook of Optical Coherence Tomography* (Bouma *et al.*, 2001). The technique has been used for the analysis of human retina (Hee *et al.*, 1995), skin and blood vessels (Barton *et al.*, 1997) and the functioning circulatory systems of small live animals (Boppart *et al.*, 1996) with excellent clarity. However, optical coherence tomography may not be suitable for the analysis of porous samples due to the multiplication of scattered photons and the resultant contribution to the signal. In order to overcome this issue, optical coherence microscopy has been developed and can be used for analysis of scaffolds that are optically opaque.

9.4.5 X-ray micro-computed tomography (MicroCT)

The X-ray micro-computed tomography (MicroCT) method is suitable for the analysis of 3D structure scaffolds, and for measuring the porosity, pore

Standardised chemical analysis and testing of biomaterials 177

Table 9.3 Abbreviations for methods

Abbreviation	Analytical method
AES	Auger electron spectroscopy including scanning tunnelling auger
AFM/SPM	Atomic force microscopy/scanning probe microscopy including topographical roughness and phase contrast
BET	Brunauer-Emmet-Teller, a porosity measurement technique
CLSM	Confocal laser scanning microscopy
DMTA	Dynamic mechanical thermal analysis
DSC	Differential scanning calorimetry
EPMA	Electron probe microanalyser
ESC	Equilibrium solvent content
EWC	Equilibrium water content
EDX-SEM	Energy dispersive X-ray analysis – scanning electron microscopy
FTIR-ATR	Fourier transform infrared spectroscopy – attenuated total reflectance
IR	Infrared spectroscopy
OM	Optical microscopy (e.g., polarised light and phase contrast)
SEM/TEM	Scanning electron microscopy/transmission electron microscopy
SPR	Surface plasmon resonance
ToF/SIMS	Time of flight/secondary ionisation mass spectroscopy
TMA	Thermal mechanical analyser
XPS/ESCA	X-ray photoelectron spectroscopy/electron spectroscopy for chemical analysis

Source: Adapted from ISO/TS, 2006.

size, shape and interconnectivity in a sample. MicroCT is a non-destructive method and does not have the limitations faced by optical techniques regarding sample thickness. In this method, the X-ray attenuation is collected at different angles when the sample is rotated, and a 3D image of a sample is generated from a series of 2D slices acquired at different heights. High-resolution MicroCT within the range of 50 nm that has the capacity to capture nano- and micro-sized pores is commercially available. This method can be used for analysis of samples in which differences between the density of solid and fluid phases is considerable. MicroCT has been used for the analysis of bone, various types of scaffolds and hydrogels.

9.4.6 Magnetic resonance imaging (MRI)

Magnetic resonance imaging (MRI) can be used for the analysis of matrices that have MR active nuclei, such as 1H, 13C. A fluid such as water is visible in MRI, since water contains MR active nuclei. However, for the analysis of a polymeric structure it is necessary to fill the pores with a low-viscosity fluid (water is again appropriate) due to the short relaxation times of nuclei on the polymer backbone. Although this method can only be used to capture pores with diameters larger than 50 µm, a technique has been developed to

178 Standardisation in cell and tissue engineering

estimate the fraction of smaller-sized pores. Due to requirement for diffusion of water into pores, this method cannot be used for measuring blind-end pores or enclosed pores.

Both MRI and MicroCT are non-destructive techniques, can be used to prepare 3D structures of opaque samples, and do not require the use of stains, dyes or any preparative method that may affect pore characteristics. Additionally, these methods can be used for monitoring cell distribution and extracellular matrix proteins within a scaffold. MRI can further be used to monitor *in vivo* samples for investigating the body's biological response to a scaffold.

9.5 Physical characterisation – permeability

Scaffold permeability is critical for cell migration. Several methods used to determine permeability are briefly described in this section.

9.5.1 Density

The pore density of a scaffold or a dried hydrogel can be easily estimated by measuring the volume (V_T) and mass (m_s) of a specimen, and knowing the density (ρ_s) of the material used to prepare the scaffold ($V_p = V_T - (m_s/\rho_s)$). The porosity percentage can be estimated by $V_p/V_T \times 100$. Care must be taken to remove any water residue to minimise errors in measuring pore density. In addition, errors in measurement can be further reduced by using the Archimedes principle to determine the volume of any specimen with irregular geometries. However, the water may not diffuse into pores of a hydrophobic specimen, and in this case, a small amount of a wetting agent (1 vol.% ethanol) or an inert gas (helium pycnometry) can be used. Methods reliant on pore density cannot differentiate between various types of pores (e.g., open, closed, and blind-end) so the data obtained from this technique may not be useful for some applications. The reliability of the data acquired by this method relies on accurate measurement of dimensions, mass and density of the sample material.

9.5.2 Porosimetry (mercury intrusion)

The total pore volume, surface area, mean pore diameter, and pore size distribution of a specimen can be estimated by this method. A non-wetting fluid, typically mercury, fills the pores in a sample, allowing the measurement of the required values. This method cannot be used to measure the porosity of soft samples such as hydrogels, as the intrusion may distort the specimen. Mercury porosimetry is unable to measure closed pores.

© Woodhead Publishing Limited, 2013

Standardised chemical analysis and testing of biomaterials   179

Gas flow porometry has only been used to measure the 'through pores' in a scaffold. In this technique, a non-volatile wetting liquid is forced into through pores by increasing gas pressure. The pore diameter measured by this method is usually smaller than that measured using other techniques. This technique is also unsuitable for soft samples such as hydrogels.

9.5.3 Nuclear magnetic resonance (NMR)

NMR's capacity to measure pore size is attributed to the fact that the behaviour of fluids in confined pores is different from their behaviour in bulk. A relatively new method, known as cryoporometry, has been used to measure pores ranging from nano-sized to 1 mm. This method is non-destructive and does not require sample preparation such as staining and dying.

9.6   Surface properties

This section has been adapted from Castner and Ratner (2002). Surface properties are one of the most critical factors affecting the biological interaction of biomaterial products. In the last two decades many advanced analytical techniques have been developed for surface analysis, which can be used to produce a better understanding about the biological reactivity of a cell with a surface. Methods such as electron spectroscopy for chemical analysis (ESCA, also called XPS), secondary ion mass spectrometry (SIMS), scanning probe microscopy and zeta potential measurement can be used for surface analysis of fragile organic surfaces, in order to avoid degradation. Additional techniques such as FTIR, Raman, sum frequency generation (SFG), and high-resolution electron energy loss (HREELS) can be used to obtain surface vibrational spectra from the surface of heat labile specimens, with minimal damage.

Commercial surface plasmon resonance (SPR) and piezo (quartz crystal) balance instruments can be used to conduct precise adsorption experiments in the nanometre thickness range, allowing assessment of the surface bio-interactions. However, the analysis of organic surfaces that are heat labile with equipment that may need to use high temperatures, such as those producing X-rays and electron beams, is still challenging.

9.6.1 Analytical techniques

Scanning probe microscopy (SPM)

SPM operates at the highest resolution, and can even be used for analysis of individual atoms. Whilst the inherent chemical specificity of this method is limited, functionalisation of the probe tip can resolve this issue. SPM can be

© Woodhead Publishing Limited, 2013

180　Standardisation in cell and tissue engineering

used to map the intermolecular forces in 3D structures, providing molecular resolutions of proteins, DNA, lipids and carbohydrates, whilst simultaneously revealing the cellular structure.

X-ray photoelectron spectroscopy (XPS) and near
X-ray absorption fine structure (NEXAFS)

The resolution of X-rays (XPS, NEXAFS, etc.) is limited in comparison to the SPM technique, due to the difficulty of focusing X-rays, but recent advances have resulted in this instrument achieving a resolution within the range of 10 nm. This method can be used for quantitative analysis of the surface composition of a specimen, and is non-destructive. However, at higher resolutions there is a risk of sample degradation due to escalating X-ray brightness (photons per unit area).

Static time-of-flight secondary ion mass spectrometry
(static ToF SIMS)

Static ToF SIMS provides detailed information about the molecular structure of organic and biological materials by creating a mass spectrum from the outer 1–2 nm of sample. The spatial resolution of this equipment is within 0.1 μm, and it can therefore be used for cellular resolution (1–100 μm). This method can be used to acquire information from the surface of biomedical specimens. The major issues in its use include the acquisition of large data sets, the production of images with low signal-to-noise ratios and the production of several peaks for identification of chemical species. The difficulty, therefore, is that whilst a large quantity of data can be obtained in a short period of time, the analysis of these data is very time-consuming. For the future development of this equipment, it is crucial that image-processing methods are enhanced to produce further details of surface chemical species quantitatively.

9.6.2　Surface analysis of hydrated samples

The analysis of wet samples can be challenging when using ultra-high vacuum equipment for surface imaging. Methods such as contact angle measurements, second harmonic generation (SFG), Brewster angle microscopy, X-ray reflectivity, SPM, frozen hydrated ultra-high vacuum (UHV) techniques (XPS, SIMS, etc.), environmental SEM, attenuated total reflectance IR (subtract out the water signal), ellipsometry, SPR and neutron scattering have been used for this purpose. These methods provide information about the composition and structure of hydrated surfaces. However, direct analysis of the sample in an aqueous medium is desirable, allowing the composition and structure to be monitored in real time.

Photon in/photon out techniques

This method is based on the concept of having longer mean free photon paths, in comparison to ion and electron paths. Photon in/photon out techniques can be used to examine the properties of solid–liquid (i.e., biomaterial in a biological environment) samples. Care must be taken when analysing samples that the bulk material contains similar species. Furthermore, it is not possible to prepare a flat surface using this method, due to loss of sensitivity.

One optical technique that can directly be used to study the structure of the solid–liquid interface is sum frequency generation (SFG). In this second-order non-linear optical process, a pulsed visible laser beam is overlapped with a tunable, pulsed IR laser beam that creates a signal at the sum frequency. Emission of the sum frequency light does not occur for the bulk phase of most materials. SFG has been used to examine the time-dependant restructuring of polymeric materials. Quantitative analysis of the concentration of surface species cannot be measured using the SFG method, which is considered a key limitation of this technique. Other photon in/photon out techniques can provide information about the thickness (ellipsometry) and amount (SPR) of a deposited species, but they are unable to give any information about the chemical composition and molecular structure. However, it is likely that the combination of these techniques may resolve the current limitations of this method for surface analysis.

9.7 Degradation and stability in physiological fluids

Degradation tests vary depending on the nature of the material, device function, application and environmental conditions (e.g., stresses, internal and external factors). Both the intended and non-intended degradation products of a material must be considered in order to evaluate its biological safety. Degradation study protocols should include analytical methods to characterise degradation products in terms of: chemical and physico-chemical properties, surface morphology, biochemical properties and causes of generation. It is critical to select a reliable, appropriate method for detecting degradation effects, which may occur in a variety of forms, including particulates, soluble compounds and ions. It is also necessary to determine the mechanism of degradation.

Degradation studies should be conducted for bioresorbable devices, which are intended to be implanted for no more than 30 days and where there is a high risk of toxic substances being released. A degradation study might not be necessary for well-established materials classified as clinically safe, the degradation of which is already thoroughly documented.

182   Standardisation in cell and tissue engineering

Degradation studies can be conducted in accelerated or real time according to ISO standards (ISO, 2009, 2012). The method of analysis relies on the nature of the material and the period that the device is expected to be kept in the body. For metallic products, for example, corrosion tests are commonly conducted. The real-time test is performed whenever the identity and quantity of degradation product is a subject of concern. A degradation test of 60 days is conducted for polymer-based devices that may be implanted for longer than 30 days, whilst a degradation period of two to seven days is common for polymeric devices that are intended to be used in body for less than 30 days. In carrying out degradation tests, care must be taken to select a representative sample, appropriate method of analysis, and pertinent choice of operating conditions for the desired application. General tests include specimen characterisation (surface area, density), solubility characterisation and microstructure assessment. In general samples, debris and fluid phases are the subject of degradation tests, and it is common to separate solid and fluid phases. Debris is dried and weighed to evaluate a change to the mass. In addition, when studying polymers, details of the molecular mass/molecular mass distribution are compared with the original material. Observation of mass loss in accelerated tests suggests the need for real-time degradation tests. Several methods have been developed for the analysis of polymeric materials (see also Table 9.4), which include:

- solution viscometry to measure molecular mass;
- swellability to assess crosslink density;
- gas and/or high-performance liquid chromatography to test for residual monomers, additives and leachables;
- size exclusion/GPC to test for molecular mass average and changes in molecular mass distribution;
- ultraviolet spectroscopy, infrared spectroscopy, NMR and mass spectroscopy to identify composition;
- atomic adsorption spectroscopy and ICP to identify catalyst and heavy metal content;
- differential scanning calorimetry to measure glass transition temperature, melting range and softening point.

Identification and quantification of degradation products from ceramics are covered by the ISO 10993-14 standard. This standard considers degradation products generated by a chemical dissociation of ceramic during *in vitro* testing, and does not cover any degradation induced by mechanical stress or external energy. Tests can be conducted in two different ways: (1) an extreme solution test conducted at low pH; and (2) in media with pH encountered in physiological fluids. The first test is typically conducted as a screen test for most ceramics.

© Woodhead Publishing Limited, 2013

Standardised chemical analysis and testing of biomaterials   183

Table 9.4 Different techniques that can be used for characterisation of biomaterials

Parameter to be analysed	Example methods (not comprehensive or exclusive)	Standard
Porosity	OM	ASTM F1854-01
Classical	Gas adsorption (BET)	ISO 18754
	Mercury porosimetry	ISO 18757
	Helium pycnometry	
Connectivity/scaffolds	SEM, AFM	
Morphology	X-ray diffraction	ASTM F665
Crystallinity	OM, DSC, SEM/TEM,	ASTM F754
	AFM	ASTM F2081
Amorphous	DMTA, AFM	ASTM F2183
Multiple phases	OM, AFM, TEM	
Hard/soft surfaces	OM, AFM/SPM, TEM	
	Ultrasound	
Surface energy/charge	Wettability (contact	
Hydrophobic/hydrophilic	angle)	
Protein adsorption/	Quartz crystal	
repulsion	microbalance (QCM) or	
	SPR, CLSM	
	Biochemical analysis	
	Radioimmunoassay	
Cell attachment/repulsion	OM, QCM, CLSM	
Abrasion resistance	IR	ASTM D968, ASTM
Stability of treated surface	Volume loss, strain	D1044 ASTM D1894,
Surface friction	gauge	ASTM D4060, ASTM
	Coefficient of friction	F732
	AFM/SPM	ASTM F735
		ASTM F1978
Topography	XPS/ESCA, ToF-SIMS	ISO 3274, ISO 4287
Surface chemical mapping	FTIR/ATR, FTIR	ISO 4288, ISO 5436-1
	microscopy	ISO 5436-2, ISO 11562
	FTIR imaging, EDX-SEM	ISO 12179, ISO 13565-1
	Raman, EPMA	ISO 13565-2, ISO
Roughness, smooth, pitted,	SEM, AFM/SPM,	13565-3
Grooved, irregular terrain	tribology,	ISO 18754, ISO 18757
('hills, valleys')	profilometry	EN 623-4
Particles	OM, laser diffraction	ASTM F1877, ISO 13319
Size, size distribution	Image analysis, filters	ISO 13320-1, ISO/TS
3D shape	(sieves), SEM	13762
		ISO 17853, EN 725-5
Swelling	EWC, ESC, image	ISO 17190-5
Water absorption,	analysis	
solvent absorption,	OM, SEM, TMA,	
shape change	microbalance	
surface crazing, weight		
gain		

Source: Adapted from ISO/TS, 2006.

© Woodhead Publishing Limited, 2013

184 Standardisation in cell and tissue engineering

9.7.1 Degradation of metallic materials

Determination of degradation products or corrosion products in metallic materials can be conducted in two different ways. The first approach assumes the use of accelerated corrosion tests as a combination of potentiodynamic and potentiostatic tests. The second is an immersion test where degradation products/released ions are assessed. Media such as artificial saliva, artificial plasma, and 0.9% sodium chloride can be selected for the degradation studies, dependent on the application of tested material.

Potentiodynamic test

This test allows potential/current density curves in the specified range (max 2 V or 1 mA/cm^2) to be recorded. Several polarisation cycles (5–10) have to be recorded and analysed to evaluate the transpassive range and breakdown potential. Potentiostatic measurements permit qualitative and quantitative determination of degradation products, which might be dissolved in the electrolyte. A test sample is kept at a constant electrode potential during the study to record the current density/time curve. The potential used to determine the degradation products should be 50 mV above the breakdown potential.

Immersion test

This test involves a sample being immersed in a glass container containing 1 mL/cm^2 of media, which is sealed to prevent evaporation and kept at physiological temperature for 7 days. The pH of the media is measured at the beginning and end of the test, and upon completion of the test, microscopic analysis is performed to observe any changes in the sample. Qualitative and quantitative analysis of electrolyte composition is performed using a method with the desired sensitivity (ppm; atomic absorption, ICP or mass spectroscopy).

9.8 Implant–tissue interface tests

The biological evaluation of medical devices is covered by the ISO 10993 standard, which outlines the principles of bio-evaluation for different types of devices based on the duration of their contact with the body, and describes a selection of appropriate tests for their assessment. Medical devices are categorised based on the nature of body contact (non-contact, surface-contacting, external communicating and implant devices) and duration of contact (limited exposure up to 24 h; prolonged exposure for 24 h to 30 days; and permanent contact for more than 30 days). Initial evaluation tests include cytotoxicity, sensitisation, irritation, intracutaneous reactivity, systemic toxicity, subchronic

© Woodhead Publishing Limited, 2013

Standardised chemical analysis and testing of biomaterials   185

toxicity, genotoxicity, implantation and haemocompatibility. Supplementary tests can include chronic toxicity, carcionogenicity, reproductive and developmental toxicity and biodegradation.

ISO standards provide clear guidance for the evaluation of the material safety and set criteria by which devices can be approved for clinical use. However, the investigation of a material often goes beyond assessing safety, further testing the efficacy of material and examining its functionality and performance in the body. These studies can include a plethora of biological, chemical, physical, biomechanical and functional tests, and the results can be directly used to back up specific items listed in the ISO standard. The literature therefore outlines several approaches and protocols for testing the relevance and reliability of materials in the biological environment. These tests are often referred to as bioactivity tests.

Wnek and Bowlin (2008) define bioactivity as the ability of a material to introduce the formation of an interfacial bonding between the implantable device and living tissues, without the formation of a fibrous capsule separating the biomaterial and the tissue. According to this definition, the term bioactivity is strictly associated with the biological reaction of a material with the surrounding environment, that is, within either the body or a simulated body environment. Bioactivity can also be defined as a process where a positive tissue response is stimulated by materials of non-biological origin when in direct contact with the tissue. More commonly, bioactivity is understood to be the induction of biologically relevant material responses, particularly on surfaces, such as the formation of apatite layers from body fluids (e.g., Kokubo, 1998; Ryhanen, 1999; Yousefpour *et al.*, 2007; Abou Neel and Knowles, 2008; Chrzanowski *et al.*, 2008a). Analysis of these responses enables the prediction of cell responses and, consequently, material integration with tissues. Such analysis allows, for example, the bone-bonding ability of materials to be predicted.

As the kinetic rate of interactions (layer formation), nature, structure and chemistry vary for different materials, the results of this test can be used to determine the bioactive differences (e.g., cell and bacterial responses) between various biomaterials. It has been demonstrated, for example, that materials that possess favourable hydroxyapatite layer formation (i.e., Ca/P ratio of 1.67) integrate more successfully in the body (Kokubo and Takadama, 2006; Abou Neel and Knowles, 2008; Chrzanowski *et al.*, 2008a, 2008b, 2012). This is most likely due to the response of cells to the presence of the naturally occurring hydroxyapatite layer.

It is important to differentiate between bioactivity and biocompatibility. The definition of bioactivity is confined to materials which in principle are approved for contact with bodily environment. There are some materials that show a high rate of apatite film formation (e.g., lead) but are not biocompatible and cannot be considered for use in biomedical applications.

© Woodhead Publishing Limited, 2013

186 Standardisation in cell and tissue engineering

The definition of biocompatibility was introduced in the early 1990s by Prof David Williams and is key to the clinical success of medical devices with low levels of immune response. Williams refers to biocompatibility as the ability of a biomaterial to perform its desired function with respect to a medical therapy, without eliciting any undesirable local or systemic effects in the recipient or beneficiary of that therapy. The biomaterial should generate the most appropriate beneficial cellular or tissue response in that specific situation and optimise the clinically relevant performance of that therapy (Williams, 2008).

Bioactivity tests are performed for all types of biomaterials and they play a critical role in the assessment and selection of materials for future *in vivo*, pre- and clinical studies. These tests are relevant for both degradable and non-degradable materials (Fujibayashi *et al.*, 2004; Kokubo and Takadama, 2006; Armitage and Chrzanowski, 2007; Abou Neel and Knowles, 2008; Byon *et al.*, 2007). Bioactivity tests are used to assess the effect of surface modification, material processing and the manufacture of new materials, composites and their blends.

Several methods have been proposed for the assessment of bioactivity (Kokubo, 1998; Chen *et al.*, 2004; Kokubo and Takadama, 2006; Cui *et al.*, 2008; Shabalovskaya *et al.*, 2008; Chrzanowski *et al.*, 2012). Such testing generally involves approaches that originated from Kokubo's laboratory, or the modified versions of these approaches proposed by Bohner and Chrzanowski (Kokubo, 2005; Bohner and Lemaitre, 2009; Chrzanowski *et al.*, 2012). These modifications have been driven by technological evolution, the development of new examination tools and deepening levels of understanding of material cell/tissue interactions. Typical variations in this methodology include the use of different simulated body fluid compositions, changes to the chemicals used, and variations in the experimental period, analytical techniques and operating conditions employed. Due to these variations, it is difficult to compare the results acquired from different research teams.

9.8.1 Tests for bone tissue implants

It is recommended that the standard method 'Implants for surgery – *in vitro* evaluation for apatite-forming ability of implant materials' (23317:2007, 2007) is used for the assessment of bone-forming ability (often referred to as bioactivity). The standard bioactivity test is conducted in several steps, which involve immersing the sample in a fluid that mimics body fluid. Initial simulated body fluids (SBFs) proposed by Hench and Kokubo (Hench, 1991; Filgueiras *et al.*, 1993; Kokubo and Takadama, 2006) lacked the organic compounds present in blood plasma, and Kokubo attempted to modify this media by addition of some organic compounds to produce a revised SBF (Kokubo, 1991; Oyane *et al.*, 2003). Following further development, a new

© Woodhead Publishing Limited, 2013

Table 9.5 The composition of the simulated body fluids (SBFs)

	Concentration (mM)							
	Na^+	K^+	Mg^{2+}	Ca^{2+}	Cl^-	HCO_3^-	HPO_4^{2-}	SO_4^{2-}
Human blood plasma	142.0	5.0	1.5	2.5	103.0	27.0	1.0	0.5
Original SBF	142.0	5.0	1.5	2.5	148.8	4.2	1.0	0
Corrected SBF	142.0	5.0	1.5	2.5	147.8	4.2	1.0	0.5
Revised SBF	142.0	5.0	1.5	2.5	103.0	27.0	1.0	0.5
New (improved) SBF	142.0	5.0	1.5	2.5	103.0	4.2	1.0	0.5

SBF was proposed to address the tendency of calcium carbonate precipitation and formation of calcite in the fluid (Oyane *et al.*, 2003). Since the first SBF was formulated several improved versions of the fluids have therefore been proposed, and the most common of these are listed in Table 9.5.

In preparing the SBF solution for immersion of the sample, it is recommended that a ratio of 1:1 volume of SBF (V_s, mL) to sample surface area (S_a, mm^2) is used, and the solution should be kept at body temperature. Samples are submerged in the fluid and can be placed in an upright (vertical) position or placed flat (horizontal) in containers with conical bottoms (e.g., Sterilin containers) to enable the homogeneous precipitation of fluid components. In the latter case, the bottom surface of the sample is examined. Samples are typically immersed for up to 4 weeks and precipitation is tested at several time intervals (e.g., 1, 3, 7, 14, 21, 28 days). The sample is taken out of solution, gently washed with pure water and dried in desiccators, so that the surface can then be examined using the thin film X-ray diffraction (TF-XRD, $2\theta = 3–50°$ for CuKα) and SEM techniques. TF-XRD enables the detection of apatite formation, while SEM images enable complementary observation of the growth of apatite crystals. However, sometimes bioactivity is estimated only on the assessment of visual film formation.

Concentrated SBF (1.5 × concentrated SBF) has been used as simulated body fluid, but it should be noted that there is no correlation between results in this solution and *in vivo* behaviour, due to the fact that current SBFs do not contain any organic compounds present in body fluid. However, this may not be a critical issue, as the formation of the apatite layer is due to the inorganic part of the media. Nevertheless, the surface interaction and adsorption of other components present in sera, such as proteins, drugs, and peptides, may have a significant impact on apatite film formation. Adsorbed proteins act as mediators between the samples and cells, which may improve cell attachment and proliferation. In contrast, protein bonding and interaction may have a negative impact on the formation of the apatite layer, due to the presence of other elements with differing charges. It is, therefore, more reliable to examine bioactivity in a modified SBF that contains both inorganic and organic compounds. Such studies are of particular interest in the study of orthopaedic biomaterials, which do not contain proteinaceous layers; in these cases proteins only come from the body environment as a result of surface adsorption.

Another issue with SBF media proposed by Kokubo for bioactivity testing is the lack of filtration of solutions and the presence of insoluble compounds, which may interfere with the formation of hydroxyapatite layers. Bohner also recommends including carbonate and CO_2 partial pressure when testing bioactivity, in order to acquire results that are closer to *in vivo* conditions (Bohner and Lemaitre, 2009).

TF-XRD cannot be used to investigate the kinetics of apatite layer formation and the effect of organic phases. Methods such as XPS, which have been developed to assess sample chemistry, offer very high sensitivity and the ability to assess thin film formation in the early stages of bioactivity testing (i.e., a few minutes after contact with media). This technique is used to observe the very first few nanometres of a surface layer, and allows very precise assessment of the chemical composition. In addition, serum protein adsorption can be investigated on the basis of nitrogen, carbon and oxygen peak evaluation. SPS can provide sufficient information to assess the rate of apatite film formation and the adsorption of organic components.

Bioactivity can therefore be used to assess osteoconduction and osteoinduction when used in conjunction with additional biological and physico-chemical tests. Bioactivity represents the potential of materials to trigger the cascade of biological reactions leading to their integration in the body or encapsulation. Typically, bioactivity is the first step enabling potential levels of osteoconductivity or osteinductivity to be predicted (Kokubo, 1998; Gu *et al.*, 2005; Garcia *et al.*, 2006; Kokubo and Takadama, 2006; Yousefpour *et al.*, 2007; Abou Neel and Knowles, 2008; Shabalovskaya *et al.*, 2008). In a classic approach the bioactivity of a sample is tested without cells, whilst osteoconductivity and osteoinductivity tests are conducted *in vitro* and *in vivo*.

9.8.2 Application specific tests

Testing the implant–bone tissue interface

As previously noted, the most common elements reviewed in analysis of these materials include material chemistry, mechanical properties, bioactivity, surface free energy (wettability), topography, cytotoxicity and degradation/corrosion properties. In the context of chemistry and mechanical properties, it is important that both bulk and surface properties are considered. Engler demonstrated that stem cells differentiate differently on substrates of different stiffness (Engler *et al.*, 2004, 2006; Rehfeldt *et al.*, 2007). It was subsequently suggested that the expression of osteoblastic markers was upregulated on substrates with Young's modules less than 34 kPa. Bone stiffness is within the range of 1–6 GPa; therefore, the stiffness required for a biomaterial suitable for use in bone implants should be within this range. Standard techniques are available for assessing the surface chemistry, topography, surface free energy and stiffness of a sample, all of which have a direct impact on cellular responses. In the context of metallic biomaterials, corrosion rates and products cannot exceed the thresholds set for each metal/alloy. It is obvious that some of the ions released from the implants can cause cytotoxic or immune responses, thus the degradation of these materials has to be at a level that does not entail any adverse reactions. Materials are therefore tested for cytotoxicity, typically using one of the following assays: MTT, MTS, alamarBlue, PrestoBlue, CellTiter, or Pico Green (Longo-Sorbello *et al.*, 2006).

By measuring metabolic activity or DNA contact, these tests are able to measure cell proliferation rates and indicate both the vitality and viability of the cells. The toxic effects can therefore be quantitatively measured by these assays. In addition, light (e.g., phase contrast) can be used to look at cell morphology, which is an indication of cells' 'wellbeing'. Cells can be fixed, dehydrated and examined by SEM, whilst electron microscopy (EDS) allows the measurement of cell chemistry, thus facilitating assessment of the minerals (Ca, P) that correlate to the differentiation and mineralisation of cells.

Testing the implant–blood interface

Tests for interactions with blood are covered by an ISO 10993-4 standard (10993-4, 2002). Devices that come into contact with blood are divided into three groups: non-contact devices (diagnostic), external communicating devices (e.g., catheters) and implant devices (e.g., stents). The type of test required depends on the level of a device's exposure to blood. Analysis of thrombosis, coagulation, platelets, haematology and the complement system are generally required, and these tests may be conducted *in vitro*, *ex vivo* and *in vivo*. It is critical to control the sterilisation of any sample that is administered in the body, particularly in contact with blood, as a lack of sterilisation can potentially cause a severe infection and a serious health risk.

Testing implant materials for pulmonary delivery

The bulk characterisation and performance properties are key determinants for the selection of materials used for pulmonary delivery. Bulk characteristics measured include particle size, morphology, density, crystallinity, charge (electrostatic), particle–particle adhesion forces and cohesiveness. As per bulk characterisation, surface morphology and roughness are assessed using SEM and atomic force microscopey (AFM). Chemistry of the materials is

190   Standardisation in cell and tissue engineering

further analysed using XPS, EDS, Raman and FTIR, whilst crystallinity is assessed using XRD, DVS and DSC. In relation to lung drug delivery, aerodynamic properties and lung deposition are examined. These tests employ various types of impactors, such as Maple-Miller, electrical low-pressure and Andersen cascade impactors (Quality Solutions for Inhaler Testing, 2012 Edition. Copley Scientific Limited; http://www.copleyscientific.com/virtual_brochure.asp?c=4).

Testing implant materials for ligament regeneration

The requirements for an artificial ligament are regulated by ISO 21534 (21534, 2007), which refers to the artificial ligament as a device intended to augment or replace the natural ligament, including any necessary fixing devices. There are a few important requirements to be considered for ligaments: intended performance (expected maximum load actions, intended minimum and maximum relative angular movement between the skeletal parts, dynamic response of the body to the shape/stiffness of the implants); design attributes (stability of the implant while allowing prescribed minimum and maximum relative movements between the skeletal parts, avoidance of cutting or abrading tissue during function other than insertion or removal, the creep resistance and rupture characteristics); materials; design evaluation (pre-clinical evaluation, clinical evaluation); manufacture and inspection; sterilisation; and packaging. In the context of the design evaluation, pre-clinical and clinical studies, the key factors to be considered are the mechanical loads and related movements to which the implants may be subjected when functioning, and fatigue testing of the highly stressed parts.

It is crucial that pre-clinical testing of implants simulates the conditions of intended use. Clinical evaluation reports should therefore be compiled, and critical analysis of relevant scientific and clinical literature covering the intended use of such implants should be conducted, including analysis of any data obtained via clinical investigation.

For any large loads in the knee region that have to be transferred in part by the ligament, fulfillment of the requirements outlined above plays a critical role in providing stability and functionality. Testing of the ligament entails very detailed characterisation of a few key areas, which can be divided into three subtests: biomechanics (static, dynamic and fatigue), chemical (structure and composition of the ligament) and biological.

Biomechanical examinations consider the strength of the construct, creeping, resistance to tearing and fatigue of the material. These tests are conducted in both physiological and non-physiological conditions. In addition, tests conducted in bioreactors are typically combined with biological testing to reveal mechanisms of remodelling and/or *de novo* tissue formation, making them of great value. Chemical and structural tests investigate the compatibility of a material in the context of its chemical composition. Structural studies link both mechanical and biological tests as they have a significant impact on both cellular responses (due to specific porosity, geometry and morphology) and mechanical properties, which are well known to be related to the structure. It is important to note that all of these aspects are highly interrelated, and unless they are all taken into account, a functional ligament will not be produced. For example, the application of load will result in stress in the ligament structure, which, in turn, will have a major impact on cell responses. As the structure of the ligament (micro- and nanostructure) provides the desired mechanical properties, these properties will also be affected by such stress.

Standards provide guidance in relation to general requirements, with essential tests reliant specifically on the particular material being studied and its intended application. Analyses should be carefully planned to comply with the set requirements. In the context of the biological evaluation, different markers relevant for ligaments (collagen, for example) can be tested. However, an artificial ligament is a gradient structure that is anchored in the insert sites (i.e., the bone channels) and passes through the osteochondral to the fibrochondral zones. As such, testing of the markers alone is not sufficient; the evaluation has to be carried out differently for each zone.

Testing implant materials for cartilage regeneration

The evaluation of materials for cartilage regeneration follows the general principles of ISO 10993 standards, regarding the structure, chemistry, biological and mechanical characteristic that should be specific to an application. *In vivo* the requirements for implantable devices intended to repair or regenerate articular cartilage are covered by ASTM F2451-05 (2010) standards. Because a high proportion of cartilage injuries in humans occur in knee joints, the knee is commonly used for the examination of cartilage repair/regeneration. It is important to consider the significant differences between animal and human models due to weight and joint anatomy, which produce significant differences in load distribution and range of motion. It has been demonstrated that mechanical loads are critical in cartilage repair, therefore the previously highlighted factors which impact on cartilage structure have to be carefully considered. In addition, the extent of compressive and shear forces in the femoral condyles, trochlear groove and tibia plateau are very different, making the site selection of critical importance. Animal studies for cartilage replacement are typically conducted on larger-sized animals, such as sheep and pigs, which have properties more closely related to those of humans, such as larger articular cartilage surfaces and greater thicknesses. In large animal models defects are typically bigger, leading to implants requiring fixation, which could have some

192 Standardisation in cell and tissue engineering

impact on the surrounding tissue. Defect type (chondral or osteochondral) and their critical size vary between 2 and 15 mm in diameter, and 1–10 mm in depth; for rabbits this is 3 mm, in sheep 7 mm and in horses 9 mm. The defect should not exceed 15–20% of articulating surface or 50–60% of the condyle width.

Another factor considered in the standard guidance is handling (activity of the animal), age (dynamic changes in metabolism remodelling and degeneration of the joint due to ageing) and study duration. For small animals 6–8 weeks is sufficient to provide information on residence time, fixation and repair progress. For large animals 8–12 weeks allows the assessment of biocompatibility and early cellular responses. Typically a period of 6–12 months is necessary to confirm the repair or regeneration of cartilage.

Test procedures include the following steps: implant preparation (nontoxic, biocompatible); defect generation; implantation and fixation; recovery and husbandry; in-life period; and necropsy. Finally, tests are evaluated based on histology, microscopic analysis and scoring, biochemistry and mechanical testing of repaired tissue.

9.8.3 Surface morphology/topography

Topography refers to the chemical mapping of a surface and the investigation of surface roughness. It is well recognised that surface topography has a major impact on biological responses. Assessment of the surface topography of medical devices is regulated by the technical specification ISO/TS 10993 (2006). For chemical mapping methods, such as x-ray photoelectron spectroscopy/electron spectroscopy for chemical analysis (XPS/ESCA), time-of-flight secondary ion mass spectrometry (ToF-SIMS), Fourier transform infrared spectroscopy/attenuated total reflectance (FTIR/ATR) and microscopy, energy-dispersive x-ray spectroscopy (EDX-SEM), Raman and electron probe micro-analyser (EPMA) are commonly used. Roughness is assessed qualitatively with SEM and profilometry, and quantitatively with SPM (e.g., AFM).

Care must be taken with regard to the scale of analysis for the topographic evaluation of a specimen, and the sensitivity of a method should also be taken into account, in case the required resolution has not been specified for measurement. The method of analysis may show a uniform composite of ceramic filler in nano-scale range, but the material may be anisotropic in the micro and/or macro scale. Therefore it is crucial to determine the size and area being analysed when reporting on the surface topography. The same is true for the selection of a test method; if we evaluate roughness using a stylus profilometer, features below the resolution achieved by the technique are not visible and cannot be detected, as shown in Fig. 9.3. However, these features could have a major impact

© Woodhead Publishing Limited, 2013

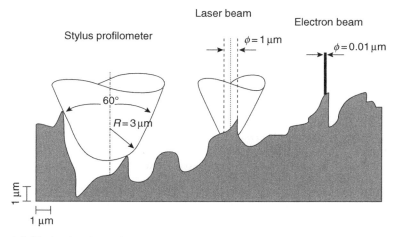

9.3 Methods of roughness evaluation.

on biological responses. Another important aspect is the presentation of roughness results, which typically use descriptions such as R_q (often assigned wrongly as RMS), R_a and R_z, which are covered by the following standards: ISO 4287 (1997), DIN 4768 and ISO 4288 (1996). It is critical to determine the cut-off length for the roughness calculations when we scan in the micro- and nano-scale ranges. The issues of resolution and post-processing are particularly relevant for scanning probe microscopy techniques.

9.9 Limitations of current standardised testing methods

Current advances in biomedical sciences have resulted in a plethora of new materials. These materials often present a different set of properties, such as size, shape and chemistry, that were not previously achievable. In addition, the properties were difficult to test in the past and technological advances have made it possible to study these materials at a new level. Currently one of the most popular areas in materials science is nanomaterials. These materials have to satisfy requirements that are regulated by standards set for the previous generation of materials. And here the question arises: are the methods proposed for materials that exist at a different scale of sub-division sensitive enough for the screening of such new materials? For example, should we use the same toxicity test for materials on the nano, micro and macro scales? In addition the sample size we can produce is often very limited and insufficient for the performance of such tests as tensile test according to the current standards. The rise of such situations has made it

194 Standardisation in cell and tissue engineering

necessary to question how materials testing should proceed: is the answer to scale up samples and perform tests accordingly? Or do we now need to look into producing updated or new standards?

9.10 References

Abou Neel, E. A. and Knowles, J. C. 2008. Physical and biocompatibility studies of novel titanium dioxide doped phosphate-based glasses for bone tissue engineering applications. *Journal of Materials Science: Materials in Medicine*, **19**, 377–386.

Armitage, D. A. and Chrzanowski, W. 2007. Surface preparation of Ni-Ti alloy using alkali, thermal treatments and spark oxidation. *Proceedings of European Society for Biomaterials 2007*.

ASTM F2150-07, A. 2007. *Standard Guide for Characterization and Testing of Biomaterial Scaffolds Used in Tissue-Engineered Medical Products*.

ASTM F2451-05. 2010. *Standard Guide for in vivo Assessment of Implantable Devices Intended to Repair or Regenerate Articular Cartilage*.

Barton, J. K. M., Thomas E., Pfefer, T. J., Nelson, J. S. and Welch, A. J. 1997. Optical low-coherence reflectometry to enhance monte Carlo modeling of skin. *Journal of Biomedical Optics*, **2**, 9.

Bohner, M. and Lemaitre, J. 2009. Can bioactivity be tested in vitro with SBF solution? *Biomaterials*, **30**, 2175–2179.

Boppart, S. A., Brezinski, M. E., Bouma, B. E., Tearney, G. J. and Fujimoto, J. G. 1996. Investigation of developing embryonic morphology using optical coherence tomography. *Developmental Biology*, **177**, 54–63.

Bouma, B. E. and Tearney, G. J. 2001. *Handbook of Optical Coherence Tomography*, New York: Marcel Dekker, Inc.

Byon, E., Jeong, Y., Takeuchi, A., Kamitakahara, M. and Ohtsuki, C. 2007. Apatite-forming ability of micro-arc plasma oxidized layer of titanium in simulated body fluids. *Surface and Coatings Technology*, **201**(9–11), 5651–5654.

Castner, D. G. and Ratner, B. D. 2002. Biomedical surface science: Foundations to frontiers. *Surface Science*, **500**, 28–60.

Chen, M. F., Yang, X. J., Hu, R. X., Cui, Z. D. and Man, H. C. 2004. Bioactive NiTi shape memory alloy used as bone bonding implants. *Materials Science and Engineering: C*, **24**, 497–502.

Chrzanowski, W., Abou Neel, E., Armitage, D. and Knowles, J. 2008a. Surface preparation of bioactive Ni–Ti alloy using alkali, thermal treatments and spark oxidation. *Journal of Materials Science: Materials in Medicine*, **19**, 1553–1557.

Chrzanowski, W., Abou Neel, E. A., Armitage, D. A. and Knowles, J. C. 2008b. Effect of surface treatment on the bioactivity of nickel-titanium. *Acta Biomaterialia*, **4**, 1969–1984.

Chrzanowski, W., Yeow, W. J., Rohanizadeh, R. and Dehghani, F. 2012. Bone bonding ability – how to measure it? *RSC Advances*, 2, 9214–9223.

Cui, Z. D., Chen, M. F., Zhang, L. Y., Hu, R. X., Zhu, S. L. and Yang, X. J. 2008. Improving the biocompatibility of NiTi alloy by chemical treatments: An in vitro evaluation in 3T3 human fibroblast cell. *Materials Science and Engineering: C*, **28**, 1117–1122.

Engler, A. J., Richert, L., Wong, J. Y., Picart, C. and Discher, D. E. 2004. Surface probe measurements of the elasticity of sectioned tissue, thin gels and polyelectrolyte

Standardised chemical analysis and testing of biomaterials   195

multilayer films: Correlations between substrate stiffness and cell adhesion. *Surface Science*, **570**, 142–154.

Engler, A. J., Sen, S. and Discher, D. E. 2006. Matrix elasticity directs stem cell differentiation. *Journal of Biomechanics*, **39**, S269–S269.

Filgueiras, M. R. T., Latorre, G. and Hench, L. L. 1993. Solution effects on the surface-reactions of 3 bioactive glass compositions. *Journal of Biomedical Materials Research*, **27**, 1485–1493.

Fujibayashi, S., Neo, M., Kim, H. M., Kokubo, T. and Nakamura, T. 2004. Osteoinduction of porous bioactive titanium metal. *Biomaterials*, **25** 443–450.

Garcia, C., Cere, S. and Duran, A. 2006. Bioactive coatings deposited on titanium alloys. *Journal of Non-Crystalline Solids*, **352**, 3488–3495.

Gu, Y. W., Tay, B. Y., Lim, C. S. and Yong, M. S. 2005. Characterization of bioactive surface oxidation layer on NiTi alloy. *Applied Surface Science*, **252**, 2038–2049.

Hee, M. R., Izatt, J. A., Swanson, E. A., Huang, D., Schuman, J. S., Lin, C. P., Puliafito, C. A. and Fujimoto, J. G. 1995. Optical coherence tomography of the human retina. *Arch Ophthalmol*, **113**, 325–332.

Hench, L. L. 1991. Bioceramics: from concept to clinic. *Journal of American Ceramic Society*, **74**, 1487–1510.

ISO 4288, I. 1996. *Geometrical Product Specifications (GPS) – Surface texture: Profile method– Rules and procedures for the assessment of surface texture*. ISO.

ISO 4287, I. 1997. *Geometrical Product Specifications (GPS) – Surface texture: Profile method – Terms, definitions and surface texture parameters*. ISO.

ISO 10993-4, I. 2002. *Biological evaluation of medical devices. Part 4: Selection of tests for interactions with blood*. ISO.

ISO 10993-19, I. T. 2006. *Biological evaluation of medical devices. Part 19 – Physico-chemical, morphological and topographical characterization of materials*. ISO.

ISO 21534, I. 2007. *Non-active surgical implants – Joint replacement implants – Particular requirements*. ISO.

ISO 23317:2007, I. 2007. *Implants for surgery – In vitro evaluation for apatite-forming ability of implant materials*.

ISO. 2009. *Biological evaluation of medical devices. Part 9: Framework for identification and quantification of potential degradation products*. ISO.

ISO. 2012. Biological evaluation of medical device. *Part 12: Sample preparation and reference materials*. ISO.

Kokubo, T. 1991. Bioactive glass-ceramics – properties and applications. *Biomaterials*, **12**, 155–163.

Kokubo, T. 1998. Apatite formation on surfaces of ceramics, metals and polymers in body environment. *Acta Materialia*, **46**, 2519–2527.

Kokubo, T. 2005. Design of bioactive bone substitutes based on biomineralization process. *Materials Science and Engineering: C*, **25**, 97–104.

Kokubo, T. and Takadama, H. 2006. How useful is SBF in predicting in vivo bone bioactivity? *Biomaterials*, **27**, 2907–2915.

Longo-Sorbello, G.S.A., Saydam, G., Banerjee, D. and Bertino, J.R. 2006. Cytotoxicity and cell growth assays. In: Julio, E.C. (ed.) *Cell Biology* (Third Edition). Burlington: Academic Press; pp. 315–324.

Oyane, A., Kim, H. M., Furuya, T., Kokubo, T., Miyazaki, T. and Nakamura, T. 2003. Preparation and assessment of revised simulated body fluids. *Journal of Biomedical Materials Research Part A*, **65A**, 188–195.

© Woodhead Publishing Limited, 2013

196 Standardisation in cell and tissue engineering

Rehfeldt, F., Engler, A. J., Eckhardt, A., Ahmed, F. and Discher, D. E. 2007. Cell responses to the mechanochemical microenvironment – Implications for regenerative medicine and drug delivery. *Advanced Drug Delivery Reviews*, **59**, 1329–1339.

Ryhanen, J. 1999. *Biocompatibility Evaluation of Nickel-Titanium Shape Memory Alloy*. OulunYliopisto: Oulu.

Shabalovskaya, S., Anderegg, J. and van Humbeeck, J. 2008. Critical overview of Nitinol surfaces and their modifications for medical applications. *Acta Biomaterialia*, **4**, 447–467.

Williams, D. F. 2008. On the mechanisms of biocompatibility. *Biomaterials*, **29**, 2941–2953.

Wnek, G. E. and Bowlin, G. L. (eds). 2008. *Encyclopedia of Biomaterials and Biomedical Engineering*: New York, London: CRC Press, Informa Healthcare.

Yousefpour, M., Afshar, A., Chen, J. and Zhang, X. 2007. Bioactive layer formation on alkaline-acid treated titanium in simulated body fluid. *Materials and Design*, **28**(7), 2154–2159.

10
Sterilisation procedures for tissue allografts

B. J. PARSONS, Leeds Metropolitan University, UK

DOI: 10.1533/9780857098726.2.197

Abstract: Terminal sterilisation of sealed packages containing healthcare products, biomaterials and tissue allografts can be the sterilisation method of choice, particularly for expensive, low production volume items such as drug-device products and biomaterials such as bone, skin, tendon and other soft matter used by tissue banks. Validation of sterility to sterility assurance levels of $1:10^6$ is based on a statistical approach and may supplement or replace other methods such as using aseptic control of the production environment. This chapter reviews the current state of the ionising radiation sterilisation protocols and their applications to a wide range of healthcare products and biomaterials.

Key words: ionising radiation, sterilisation, international standards, biomaterials.

Note: This chapter has been previously published in *Sterilisation of biomaterials and medical devices,* eds S. Lerouge and A. Simmons, Woodhead 2012, under the title 'Sterilisation of healthcare products by ionising radiation: principles and standards' by B. J. Parsons.

10.1 Introduction

A wide range of products is included under the heading of healthcare products, such as syringes, catheters, dressings, sutures, tissue allografts, proteins, enzymes, drugs, polysaccharides, liposomes and bones. They are also used in combination with other components, such as metals and polymers, to produce drug-device combination products, an area of both rapid development and growth in the pharmaceutical industry. Pharmaceuticals, including drugs and devices, are sterilised using a range of techniques, including dry heat, ethylene oxide, hydrogen peroxide, air-steam mixtures, steam, steam-in-place, gas plasma, filtration and formaldehyde and ionising radiation (Agalloco and Akers, 1993; Nordhauser *et al.,* 1998).

The choice of sterilisation technique will depend upon many factors, particularly the desired level of sterility, applicability to both large- and small-scale production facilities, validation of the process and potential damage to the healthcare product. The use of large-scale facilities to sterilise

198   Standardisation in cell and tissue engineering

small production runs of expensive items such as drug-device combination products is unlikely to be cost-effective and is also difficult to validate with regard to the sterility assurance level (SAL).

The attraction of using ionising radiation for the sterilisation of a sealed package containing a healthcare product is clear and this approach is now widely used to sterilise mass-produced items, such as medical syringes, sutures, needles and dressings, where ionising radiation damage is either unlikely or has little effect on the effectiveness and safety of the product. International standards, particularly applicable to mass-produced items, are now available to ensure the effectiveness of terminal sterilisation of healthcare products by ionising radiation, typically at sterility assurance levels of $1:10^6$ (ISO, 1995a, 1995b, 1996, 2006a, 2006b, 2006c; AAMI, 2001). These standards have also formed the basis of a code of practice (Parsons *et al.*, 2005; IAEA, 2008) for the terminal sterilisation of tissue allografts, where other considerations need to be taken into account in the case of items such as bone and amnion that are diverse both in origin and nature.

The use of ionising radiation to sterilise healthcare products is particularly attractive for many applications. Terminal sterilisation of relatively clean products in a sealed package combined with a statistical approach to dose setting to achieve a desired SAL are the major advantages of this technique. It can be applied to both large- and small-scale production runs with relatively easy and demonstrable validation procedures. As with other sterilisation techniques, damage to the healthcare product, particularly to sensitive products such as proteins, enzymes and drugs, must be minimised and be constrained within acceptable limits. The aim of all sterilisation processes is to reduce bacterial and viral contamination to acceptable levels while retaining the integrity and functionality of the product. In order to devise suitable sterilisation processes using radiation, it is therefore essential to understand the principles of radiation chemistry and how sterilisation processes using ionising radiation can be validated to ensure they meet these objectives.

10.2   Interaction of ionising radiation with matter

The main sources of radiation used to sterilise biomaterials are: (a) high-energy photon sources such as X-ray machines and cobalt-60; and (b) high-energy electrons from electron accelerators. Both types ionise molecules but via different processes, which affect their practical application to sterilisation.

10.2.1   The ionisation of molecules

If the energy of a particle or photon exceeds the ionisation potential of a molecule, then, in principle, ionisation may occur. In practice, sources of

© Woodhead Publishing Limited, 2013

ionising radiation have energies which greatly exceed the ionisation potentials of all molecules and are usually classified by the way in which they are produced. For the purposes of sterilisation of healthcare products, both high-energy photon and high-energy electron sources are used commercially. High-energy photons and high-energy electrons interact with matter in substantially different ways and these are outlined below.

10.2.2 High-energy photons

High-energy photons, produced by either X-ray machines or by gamma irradiators, interact with matter in three distinct processes: via the photoelectric effect, via Compton and other scattering processes and via a pair formation process. The contribution of each process to the absorption of photons depends on both the energy of the photons and the atomic number of the stopping matter.

For low-energy photons interacting with water, the photoelectric effect is dominant at 0.01 MeV and tails off at 0.1 MeV. In this process, the photon is completely absorbed by the water molecules and a photoelectron is ejected. At higher energies, Compton scattering is the dominant process in water over a wide range of photon energies (approximately 0.1–10 MeV). In Compton scattering, only a fraction of the photon energy is absorbed to produce an ejected electron and so the degraded, scattered photon continues to ionise more water molecules. At high energies in excess of twice the rest mass of the electron – that is, in excess of 1.02 MeV – the incident photons can also be absorbed and in doing so produce a positron and electron pair.

As a beam of high-energy photons with incident intensity I_0 passes through matter, the loss of intensity may be calculated using the Beer-Lambert equation:

$$I = I_0 e^{-(\mu/\rho)x\rho}$$

where x denotes the path length (cm), μ/ρ is the total mass attenuation coefficient (cm^2 g^{-1}) and ρ (g cm^{-3}) is the density of the matter. In this equation, the total mass attenuation coefficient takes into account the contributions from all three photon absorption processes – that is, from the photoelectric effect, Compton scattering and pair formation. Calculations of the total mass absorption coefficients for a wide range of photon energies absorbed by specific atoms, compounds and mixtures are available, for example, from the National Institute of Standards and Technology (Hubbell, 1977, 1985). Figure 10.1 shows a plot derived from such calculations for water.

Applying values of μ/ρ for water taken from Fig. 10.1 of 6.323×10^{-2} cm^2 g^{-1} at 1.25 MeV (as an approximate value for the 1.173 and 1.332 MeV

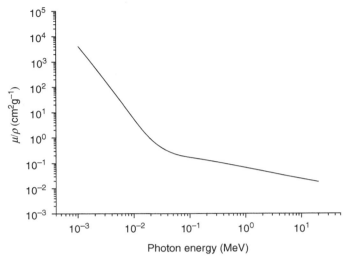

10.1 Mass attenuation coefficients (μ/ρ; cm^2 g^{-1}) for water for high-energy photons (Hubbell, 1977, 1985).

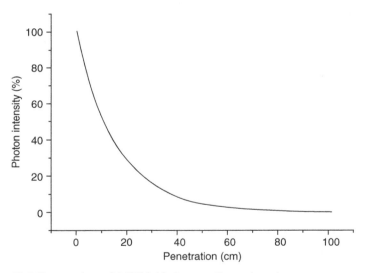

10.2 Penetration of 1.25 MeV photons through water.

gamma rays emitted by ^{60}Co), Fig. 10.2 shows the effect of depth of water ($\rho = 1$ g cm^3) on the intensity (I) of these gamma rays. It is clear, therefore, from Fig. 10.2 that high-energy photons are highly penetrating and can provide well-distributed and uniform sterilisation doses of radiation

Sterilisation procedures for tissue allografts 201

to large packages of healthcare products. It is important to note that high-energy photons lose their intensity exponentially and, therefore, unlike high-energy electrons, they do not have a finite range as they pass through matter.

10.2.3 High-energy electrons

High-energy electrons also cause ionisations in atoms and molecules as they pass through matter. The mechanism by which they lose energy is, however, different to those involved in the loss of energy by photons (see Section 10.2.2). The loss of energy as a function of distance travelled through matter, denoted by the term 'stopping power', is described by the Bethe equation in which the rate of change of energy loss with distance depends both on the energy of the electron and on the electron density of the stopping matter. This equation shows that the stopping power increases as the electron energy decreases. A consequence of this is that electrons deposit more energy per unit distance as they slow down and thus have a finite range. High-energy electrons are thus much less penetrating than high-energy photons and produce much denser ionisation within matter, with the secondary electrons produced by the primary ionisation process being produced in a cascade and having sufficient energy to bring about many more ionisations. The radiation dose thus varies with penetration depth in a characteristic way with the maximum dose being dependent upon electron energy and always occurring between the incident surface and the range of the electron. In water, for example, a 10 MeV electron will penetrate only about 5.2 cm, although the absorbed radiation (usually expressed in Grays (Gy) where $1 \, Gy = 1 \, J \, kg^{-1}$) will be relatively small beyond about 4 cm. This is illustrated in Fig. 10.3.

High-energy electrons can also produce high-energy photons in the form of characteristic X-rays and bremsstrahlung radiation. The probability of producing these photons is dependent upon the atomic number of the matter absorbing the electrons and is significant for heavy metals such as tungsten and molybdenum. Characteristic X-rays are produced when incoming electrons cause the ejection of electrons from the inner atomic shells of the heavy metal target – the X-rays are emitted when other higher-energy electrons within the metal occupy the vacancies produced by the initial electron ejection processes. These X-rays have narrow bands of energy and are characteristic of the heavy metal target. Bremsstrahlung radiation, which has a continuum of energy from zero up to that of the incident electrons, is produced when the incident electrons interact with the electric field of the heavy metal nucleus. The maximum intensity of bremsstrahlung radiation occurs at approximately one-third of the maximum energy, the energy of the incident electrons.

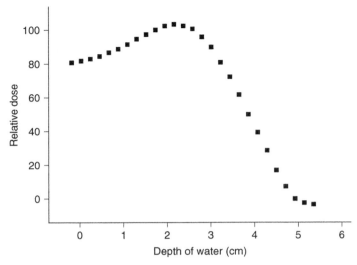

10.3 Penetration of 10 MeV electrons through water. *Source*: Data supplied by Andrew Stirling, I-Ax Technologies, Inc., Ottawa, Canada.

10.3 Sources of ionising radiation

Both high-energy photon sources and high-energy electron accelerators are in common use for the sterilisation of biomaterials. The choice of sterilisation source will depend upon a number of factors which are discussed in the following section.

10.3.1 Gamma radiation and X-ray sources

Cobalt-60 (^{60}Co) and caesium-137 (^{137}Cs) are the most widely used sources of gamma radiation. ^{60}Co produces gamma rays with energies of 1.173 and 1.332 MeV and has a half-life of 5.27 years, whereas ^{137}Cs produces gamma rays with an energy of 0.662 MeV and has a longer half-life of 30.1 years. For both isotopes, the gamma rays' energies are not high enough to induce radioactivity in the irradiated products, which would otherwise be a serious disadvantage to a sterilisation process. In industrial practice, the use of ^{137}Cs has been limited to small self-contained, dry storage irradiators used primarily for the irradiation of blood and for insect sterilisation.

In principle, X-rays may also be used for sterilisation. For example, high-energy electrons produced by an accelerator could be used to produce high-energy photons (e.g., X-rays produced by bombarding a tungsten target). In practice, however, the costs of establishing and running such a facility are relatively high with only low conversion of electron beam power to X-ray beam power (I-Ax Technologies Inc., 2008). The use of high-energy electron sources to produce high-energy X-rays is also limited by the potential for producing radioactivity in the irradiated product. This can occur via a number of processes, including photodisintegration, neutron activation and photoactivation. However, extensive research has shown that below certain energy thresholds, any induced radioactivity is insignificant compared with that which is naturally present. These limits have been agreed on by the Joint Expert Committee on Irradiated Foods of the UN Food and Agriculture Organization, the World Health Organization and the International Atomic Energy Agency. They have also been accepted by the US Food and Drug Administration and other national bodies. The limits are currently 10 MeV for electrons and 7.5 MeV for high-energy photons.

10.3.2 High-energy electron sources

High-energy electrons are produced industrially using electron accelerators. Two types are in common use: DC accelerators and accelerators based on radio frequency (RF) power technology. For the former type, the most commonly commercially available types are the Dynamitron® and the Insulated Core Transformer supplying up to 5 MeV electrons. Higher-energy electrons are produced by the RF accelerators, easily reaching energies of 10 MeV. Radio frequency accelerators, which use a series of RF cavities, are called linacs and can either be S-band (operating at an RF of 3 GHz) or L-band (1 GHz). S-band accelerators produce beam powers up to 20 kW, whereas L-band machines can produce beam powers in excess of 20 kW. A more compact RF accelerator, the Rhodotron®, uses radial accelerating fields.

10.3.3 Commercial radiation sources

Since the mid-1950s, there has been a rapid growth in the use of ionising radiation to sterilise or reduce the microbial bioburden of a range of industrial and agricultural products. The growth was largely stimulated by the need for single-use medical devices. Both gamma ray sources (^{60}Co) and electron beam sources are used currently for industrial sterilisation, and have been developed either as 'in-house' facilities or as outsourced contract services. Until 2006, about 65% of sterilisation activity was provided by outsourced contract service providers (Masefield et al., 2006).

In a worldwide survey of ^{60}Co radiation sources made by the International Atomic Energy Agency (IAEA), 123 radiation-processing facilities were listed in a directory (IAEA, 2004). Of these, 104 were used for the sterilisation of healthcare products such as medical devices, biological tissues and sanitary materials at a rate of 336 000 m^3 per annum. In a separate survey of

204   Standardisation in cell and tissue engineering

industrial electron beam facilities made by I-Ax Technologies Inc., a total of 42 facilities were being used for sterilisation purposes (14 in North America, 20 in Europe and 8 in Asia) (I-Ax Technologies Inc., 2008). The above data reflect the current preference for a gamma radiation source for sterilising a wide range of products. This choice of source is seen to provide a flexible, versatile and cost-effective method of sterilisation. However, for low-density products with a uniform composition and compact packaging, electron beam accelerators can provide much faster processing. X-ray sources are emerging as an industrial method of sterilisation. In 2008 IBA and LEONI Studer jointly announced the construction of a new X-ray sterilisation facility using the latest Rhodotron® TT-1000 system at the LEONI Studer Hard premises in Däniken, Switzerland. This will be the first facility worldwide capable of sterilising large amounts of medical devices directly on pallets, using an X-ray system.

10.4   Validation and international standards of sterilisation by ionising radiation

As with all methods of sterilisation, it is imperative that internationally accepted validation processes are adopted. These protocols may vary in different parts of the world with major protocols usually being developed in Europe and the USA. The following section outlines the principles involved in such protocols for sterilisation by irradiation.

10.4.1   Principles

The main aim of sterilisation of healthcare and related products is to reduce the level of pathogens to an acceptable, safe level. In doing so, it is clearly important to minimise damage to the product itself. The principles of radiation chemistry and the methods derived therefrom have been summarised above.

The action of radiation on bacteria, viruses and spores has received much attention in the research literature, largely, in the case of cells, as part of the process of understanding the mechanisms of radiotherapy. Cells, including bacterial cells, are killed by ionising radiation through damage to DNA. The damage may be attributable to both the indirect effect, arising from water-derived free radicals produced within the cell, and also the direct effect of radiation on DNA within the cell nucleus. It is unlikely that water-derived free radicals formed outside the cell are lethal. The proportions of indirect to direct effect within cells and viruses have also been the subject of much study and estimates have been made which are close to 50:50 (e.g., see von Sonntag, 1987; Krisch et al., 1991). That the lethal effects of radiation occur

© Woodhead Publishing Limited, 2013

Sterilisation procedures for tissue allografts   205

within the cell or virus is a distinct advantage for sterilisation by radiation of healthcare products in solution. Thus, water-derived free radicals produced outside the cells or viruses can be scavenged. At the same time, the direct effect of radiation on DNA still occurs and so kills cells. There may be some effect of free radical scavengers incorporated within the cell which might reduce the rate of killing and this can be tested and accounted for. The effect of absorbed dose on the inactivation of a population of a specific cell or virus is normally accounted for quantitatively by an exponential relationship:

$$N = N_0 10^{-(D/D_{10})}$$

where N represents the number of survivors at a dose D, N_0 is the original number of cells or viruses and D_{10} is the dose required to reduce the number of cells or viruses to 10%. Differences in sizes of the genomes for bacteria, spores and viruses lead to differences in sensitivity to radiation. In general, D_{10} values for bacteria and spores are lower than those for viruses. Typical D_{10} values for bacteria, for example, range from 1 to 4 kGy whereas the typical range for viruses is about 3–8 kGy. It should be assumed that D_{10} values are temperature-dependent. For the HIV-1 virus, for example, the D_{10} value was found to be 7.2 kGy at room temperature and 8.3 kGy at $-80°C$ (Hernigou *et al.*, 2000). Other factors may also affect the D_{10} values and it is therefore advisable to determine these values for the particular set of sterilisation conditions.

10.4.2 International standards

The quantitative relationship between cell or virus survival and also the ability of commercial gamma radiation sources to deliver accurate doses of radiation enables methods to be developed which can achieve specific SAL for healthcare products. These methods were first developed for manufactured healthcare products, such as syringes, sutures, needles and items produced in large numbers. In the seminal ISO (International Organization for Standardization) documents on these methods (ISO, 1995a, 1995b), two methods were used to establish radiation doses to achieve SAL values of 10^{-6} (i.e., a probability of 1 in 10^6 of there being one survival colony-forming unit (CFU) in the case of bacteria). Method 1 relied on knowing the bioburden on the product before irradiation – that is, the CFU values for each type of bacteria should be known. In this method, for bacteria, a standard distribution of resistance (SDR) was assumed, as given in Table 10.1. These data were then used to establish a verification dose to achieve a SAL of 10^{-2}.

Delivery of the verification dose and subsequent confirmation that no CFUs survive allowed a sterilisation dose to achieve an SAL of 10^{-6} to be

© Woodhead Publishing Limited, 2013

206 Standardisation in cell and tissue engineering

Table 10.1 Microbial standard distribution of resistance

D_{10} (kGy)	1.0	1.5	2.0	2.5	2.8	3.1	3.4	3.7	4.0	4.2
%		65.487	22.493	6.302	3.179	1.213	0.786	0.350	0.111	0.072	0.007

Source: Whitby and Gelda (1979).

calculated. The method involved a statistical approach to setting the sterilisation dose, requiring the use of relatively large numbers of samples from three batches (130) for the establishment of the initial bioburden and verification dose. In Method 2, no assumptions were required concerning the numbers and types of bioburden. Instead, incremental doses were given to samples of the product and the remaining survivors measured. Again, a relatively large number of samples (280) was required to establish a verification dose for an SAL of 10^{-2}, from which the sterilisation dose required to achieve an SAL of 10^{-6} could be calculated.

These seminal international standards ISO 11137:1995 (ISO, 1995b) have now been cancelled and replaced by ISO 11137, Parts 1–3, 2006 (ISO, 2006a, 2006b, 2006c) thereby allowing revisions of the methods 1 and 2 and also inclusion of a new method, the VD_{max} method. Part 1 of the revised standards in which the requirements for development, validation and routine control of a sterilisation process for medical devices, are set out, is designed to ensure that the activities associated with the process of radiation sterilisation are performed to the required standard. These activities include calibration, maintenance, product definition, process definition, installation qualification, operational qualification and performance qualification. It also emphasises that attention should be given to other aspects of the whole sterilisation process, from raw material to the final sealed, sterilised product package. Such considerations include:

- the microbiological status of raw materials;
- the validation and routine control of any cleaning and disinfection procedures used on the product;
- the control of the environment in which the product is manufactured, assembled and packaged;
- the control of equipment and processes;
- the control of personnel and their hygiene;
- the manner and materials in which the product is packaged; and
- the conditions under which the product is stored.

Part 3 of ISO 11137 (ISO 11137, 2006c) gives guidance on how the dosimetric requirements of the ISO should be met. The measurement of dose is central to the sterilisation process. An accurate and precise dose delivered to a product whose initial bioburden is known enables a statistical approach to be taken for the achievement of a specified SAL, the central feature of

© Woodhead Publishing Limited, 2013

Sterilisation procedures for tissue allografts   207

the sterilisation of medical products by ionising radiation. Radiation dose is measured during all stages of development, validation and routine monitoring of the sterilisation process. It is important to demonstrate that dose measurement can be related to an international standard, that the uncertainty of measurement is known and that the influence of temperature, humidity and other environmental considerations on dosimeter response is known and taken into account.

Dose mapping is a particularly important parameter in the determination of the uncertainty of the dose delivered to products. The mapping process is essentially a measurement of the variation of delivered dose within the radiation containers in which products are irradiated. The variation of dose can be influenced by the density of products where a low-density product would not significantly shield dose from other products within the container. Other considerations include the size of products and their spatial arrangement within the radiation container. Dose mapping considerations also vary according to the type of irradiation facility: gamma, electron beam or X-ray. The main outcome of a thorough and proper consideration of dosimetry in a sterilisation process is the establishment of minimum and maximum doses delivered to containers. These limits are clearly important for establishing doses which will guarantee sterilisation of any sealed product package to the specified SAL.

In Part 2 of ISO 11137 (ISO 11137, 2006b), the methods for establishing the sterilisation dose have been both revised and amended to include an approach based on the VD_{max} method (AAMI, 2001). This document also introduces the concept of product families, the grouping of which is largely dependent upon the number and types of microorganism present on or in the product. The criteria for including a product within a product family also includes other parameters which may affect bioburden, such as the nature and sources of raw materials, the components, the product design and size and the manufacturing process, equipment, environment and location. The ISO also categorises the types of manufactured items that can be sterilised. These are: individual healthcare products in their packaging systems; a set of products within a packaging system to form a healthcare product; a number of identical healthcare products in their packaging system; and a kit comprising a variety of procedure-related healthcare products. Guidance on the selection of items within these categories is then given for the purposes of dose setting and dose substantiation.

An important aspect of establishing the sterilisation dose is the decision whether to test whole individual products within the above product categories, or to test instead a sample item portion (SIP). The latter may be taken when it is otherwise impracticable to test the whole product, providing the average bioburden of the individual product is greater than 1 CFU. The value of the SIP is a fraction whose value can be calculated using the

208   Standardisation in cell and tissue engineering

ISO guidelines. Thus, for a powder, its mass can be used. The adequacy of the SIP must be demonstrated – out of 20 non-irradiated SIPs, at least 17 should yield positive tests of sterility – that is, 17 should show detectable microbial growth. Tests of sterility should be conducted in accordance with ISO 11737-1 and ISO 11737-2.

In Part 2 of ISO 11137, a number of approaches to setting the sterilisation dose are given, allowing the manufacturer of healthcare products a considerable degree of flexibility to achieve the desired SAL. Essentially, there are three methods: Method 1, Method 2 and VD_{max} methods.

In Method 1, the determination of the sterilisation dose depends on experimental verification that the radiation resistance of the product bioburden is less than or equal to the resistance of a standard distribution of resistances, as detailed in Table 10.1 (Whitby and Gelda, 1979). The method is based upon determining the initial bioburden of the unirradiated product using at least ten product items from each of three independent production batches. Using an appropriate average bioburden, the dose required to yield an SAL of 10^{-2} can be calculated using an extended form of the single exponential D_{10} equation given above applied to the SDR in Table 10.1. Tables of radiation doses for combinations of bioburden and SAL for the SDR are given in ISO 11137-2 for a wide range of bioburden values for each of the SAL values in the range 10^{-2}–10^{-6}. The verification dose test to achieve an SAL of 10^{-2} requires, thus, 100 product items from a single batch to be irradiated and sterility tests carried out on them. The verification dose test may be accepted if there are no more than two positive tests of sterility within the 100 items, and provided that the actual dose or range of doses delivered to the items are within limits set out in the ISO. The sterilisation dose required, for example, to achieve an SAL of 10^{-6} can then be calculated using the appropriate tables in ISO 11137-2. If SIPs are used, the average bioburden for the whole product must first be calculated before using the tables to calculate the sterilisation dose. This version of Method 1 applies to products with an average bioburden ≥1.0 for multiple production batches. ISO 11137-2:2006 also gives amendments to this procedure for both single production batches (bioburden ≥1.0) and for products with average bioburden in the range 0.1–0.9 for either single or multiple batches.

In Method 2, no assumptions are made about the radiation resistance of contaminating microorganisms and there are no requirements to measure the initial bioburden. Instead, incremental doses are given to a number of products in order to estimate the dose at which only one in 100 products would be expected to be non-sterile. At this dose, the D_{10} value (the dose required to reduce the number of microorganisms to 10% of this value) for the remaining microorganisms should be more homogeneous, and it is this value which is then used to calculate the dose to achieve a higher level of sterility assurance, typically 1 in 10^6. At each incremental dose, the number

© Woodhead Publishing Limited, 2013

Sterilisation procedures for tissue allografts   209

of positive tests of sterility is recorded, this number decreasing as the dose is increased. There are two variations of Method 2: 2A and 2B. The former is used more generally while the latter is used for products with a low and consistent bioburden.

In Method 2A, 280 product items are selected from each of the three independent production batches. From each production batch so selected, 20 product items are irradiated at each of at least nine doses starting at 2 kGy and increasing the dose by 2 kGy increments. For each product item, the number of positive tests of sterility is recorded and then this information is used to determine the dose to provide an SAL of 10^{-2} for the test. It is this dose which is subsequently used in a verification dose experiment on a further 100 product items. In this latter test, the number of positive tests of sterility is recorded and, depending upon the number found in the range 0–15, a sterilisation dose to achieve an SAL of 10^{-6} may be calculated. If the number of positive sterility tests exceeds 15, the cause should be determined and corrective action implemented before a new determination of a sterilisation dose can take place.

Method 2B is similar to 2A in that incremental doses are again used to test the actual radiation resistances on the products. In this case, however, the entire product should be used (SIP = 1), the number of positive tests of sterility should not exceed 14 in the incremental dose tests and the estimate of the dose required to produce an SAL of 10^{-2} should not exceed 5.5 kGy.

In the third method of ISO 11137, Part 2 (ISO 11137, 2006b), the VD_{max} method, there are similarities with Method 1, in that the initial bioburden values on product samples are required and the SDR is the basis of the assumed radiation resistance. By taking into account the distribution of radiation resistances in the SDR, a verification dose experiment carried out on only ten product items is calculated – that is, for an SAL of 10^{-1} which is characteristic of both bioburden level and the associated maximal resistance. Thus, components of the SDR of high D_{10} value are used to determine the sterilisation dose and so ensure that a greater degree of conservativeness of the SDR is preserved. The method is designed not only to provide this degree of assurance but also to facilitate the use of fewer product items for testing – in this case, only ten items. In practice, the VD_{max} dose is calculated using the average bioburden level and then ten product items from each of the three independent production batches are exposed to this dose and each item subjected to a test of sterility. If there is no more than one positive test of sterility in the ten tests, the pre-selected sterilisation dose is substantiated. The pre-selected sterilisation doses are 15 and 25 kGy, the former applicable only for product bioburdens less than or equal to 1.5, while the latter is applicable to average bioburden levels of less than or equal to 1000. Modifications of the VD_{max} methods may also be used for items from a single production batch.

© Woodhead Publishing Limited, 2013

210 Standardisation in cell and tissue engineering

10.5 Conclusions and future trends

There is an increasing realisation that the established standard sterilisation dose of 25 kGy confers no guarantee of sterility. This is particularly so for viral contamination where the D_{10} values are generally higher than for bacteria. Good manufacturing practice, where the statistically averaged bioburden is low and the distribution of pathogens is known in both types and intensities, is likely to allow the use of doses lower than 25 kGy to achieve sterility assurance levels of 1:10^6. Recent revisions of the ISO standards (ISO, 2006a, 2006b, 2006c) facilitate the use of both lower doses of sterilisation and fewer samples for dose validation purposes, while retaining the integrity of the approach to attain sterility assurance levels as high as 1:10^6. These and future revisions are likely to be of particular interest and significance to the drug-device industry where low volumes of expensive products present a challenge to the cost-effectiveness of sterilisation processes designed for much larger product volumes.

10.6 Sources of further information and advice

The Chemical Basis of Radiation Biology by von Sonntag (1987) is recommended for insight into the mechanistic aspects of the radiation chemistry and biochemistry of cells, viruses and their components. For guidance on standards to be adopted for the sterilisation of healthcare and tissue allografts, the ISO and IAEA references (and revisions thereof) contained in this chapter are essential reading. Finally, the websites of the International Irradiation Association and of the International Atomic Energy Agency are very useful to keep abreast of current practices in sterilisation and for international meetings relevant to sterilisation by ionising radiation.

10.7 References

Agalloco, J. and Akers, J. 1993. Validation of sterilization processes and sterile products. In Avis, K. E., Leberman, H. A. and Lachman, L. (eds), *Pharmaceutical Dosage Forms: Parenteral Medications*. London: Informa Health Care, 231–287.

American Association of Medical Instruments (AAMI). 2001. Sterilization of health care products – radiation sterilization – substantiation of 25 kGy as a sterilization dose, *AAMI TIR,* Arlington, VA.

Hernigou, P., Gras, G., Marinello, G. and Dormant, D. 2000. Influence of irradiation on the risk of transmission of HIV in bone grafts obtained from appropriately screened donors and followed by radiation sterilization. *Cell and Tissue Banking,* **1,** 279–289.

Hubbell, J. H. 1977. Photon mass attenuation and mass energy-absorption coefficients for H, C, N, O, Ar and 7 mixtures from 0.1 keV to 20 keV. *Radiation Research,* **70,** 58–81.

© Woodhead Publishing Limited, 2013

Sterilisation procedures for tissue allografts 211

Hubbell, J. H. 1985. Photon cross-sections 1 keV to 100 GeV – current NBS compilation. *Transactions of the American Nuclear Society*, **50**, 153–154.

IAEA. 2004. *Directory of Gamma Radiation Processing Facilities in Member States*. IAEADGPF/CD (ISBN 92-0-100204), Vienna.

IAEA. 2008. *Radiation Sterilization of Tissue Allografts: Requirements for Validation and Routine Control. A Code of Practice*. Vienna: International Atomic Energy Agency.

I-Ax Technologies Inc. 2008. *Application Note: 900*. Iaxtech.com.

ISO. 1995a. Sterilization of medical devices – microbiological methods – Part 1: Estimation of population of microorganisms on products. ISO, 11737-1. Geneva: International Organization for Standardization.

ISO. 1995b. Sterilization of health care products – requirement for validation and routine control – radiation sterilization. ISO, 11137. Geneva: International Organization for Standardization.

ISO. 1996. Sterilization of health care products – radiation sterilization – substantiation of 25 kGy as a sterilization dose for small or infrequent product batches. ISO/TR, 13409. Geneva: International Organization for Standardization.

ISO. 2006a. Sterilization of health care products – radiation – Part 1: Requirements for development, validation and routine control of a sterilization process for medical devices. ISO, 11137-1. Geneva: International Organization for Standardization.

ISO. 2006b. Sterilization of health care products – radiation – Part 2: Establishing the sterilization dose. ISO, 11137-2. Geneva: International Organization for Standardization.

ISO. 2006c. Sterilization of health care products – radiation – Part 3: Guidance on dosimetric aspects. ISO, 11137-3. Geneva: International Organization 2012/2013 for Standardization.

Krisch, R. E., Flick, M. B. and Trumbore, C. N. 1991. Radiation chemical mechanisms of single strand and double strand break formation in irradiated SV40 DNA. *Radiation Research*, **126**, 251–259.

Masefield, J., Liu, D. and Brinston, R. 2006. *The Future of Radiation Processing, Medical Device Development*. London: SPG Media Ltd, 13–15.

Nordhauser, N. M., Nordhauser, F. M. and Olson, W. P. 1998. *Sterilization of Drugs and Devices: Technologies for the 21st Century*. Illinois, USA: CRC Press.

Parsons, B. J., Kairiyama, E. and Phillips, G. O. 2005. The development of a Code of Practice for the radiation sterilization of tissue allografts. In Kennedy, J. F., Phillips, G. O. and Williams, P. A. (eds), *Sterilization of Tissues Using Ionizing Radiations*. Cambridge: CRC Woodhead, 39–63.

von Sonntag, C. 1987. *The Chemical Basis of Radiation Biology*. London: Taylor & Francis.

Whitby, J. L. and Gelda, A. K. 1979. Use of incremental doses of cobalt 60 radiation as a means to determine radiation sterilization dose. *Journal of the Parenteral Drug Association*, **33**, 144–155.

© Woodhead Publishing Limited, 2013

11

Commercial manufacture of cell therapies

I. B. WALL, University College London, UK and
D. A. BRINDLEY*, University of Oxford, UK and
Harvard University, USA

DOI: 10.1533/9780857098726.2.212

Abstract: The commercial manufacture of cell-based therapies is expected to revolutionise healthcare provision over the coming decades and address growing clinical needs of an aging population and associated age-related degenerative disorders. Exciting advances in the emerging field of stem cell biology have great potential to treat many of these diseases. This chapter provides an overview of the challenges of moving from a laboratory-scale experiment to regulatorily sound industrial manufacturing platforms. The adoption of standardised protocols to attain a clearly defined, high-quality product across multiple batches is essential, as is the capacity for scale-up processing for allogeneic cell therapy or scale-out processing for autologous therapy.

Key words: cell therapy, commercialisation, scale-up, clinical trials, manufacturing, bottlenecks, regenerative medicine, platform process.

11.1 Introduction: cells as therapies

The concept of 'cells for therapy' is not new. Bone marrow transplants have been successfully conducted since the 1950s (Thomas *et al.*, 1957) and advances in stem cell biology over recent years have led to the emergence of the regenerative medicine sector. Cell-based therapies are a very important

---

* Disclosure: D.A.B. has no other relevant affiliations or financial involvement with any organisation or entity with a financial interest in or financial conflict with the subject matter or materials discussed in any postings apart from those disclosed. D.A.B. is subject to the Chartered Financial Analyst (CFA) Institute's Codes, Standards, and Guidelines, and as such, the author must stress that his contributions are provided for academic interest only and must not be construed in any way as an investment recommendation. Further, D.A.B. is a stockholder in Translation Ventures Ltd, a company that, amongst other services, provides commercial manufacturing advice to clients in the cell therapy sector. At the time of writing, D.A.B., within the last seven years, has received and/or pending consultancy and/or funding to the value of greater than $10000 from the following relevant organisations: Wellcome Trust, Technology Strategy Board, GE Healthcare, TAP Biosystems, Nature Publishing Group, UK Cell Therapy Catapult, Centre for the Commercialization of Regenerative Medicine, California Institute for Regenerative Medicine, SENS Research Foundation, Lonza, NIH – and Athersys Inc, Lawford Davies Denoon and the Regulatory Affairs Professional Society to the value of less than $10000.

212

© Woodhead Publishing Limited, 2013

Commercial manufacture of cell therapies

part of this sector and are gaining considerable interest as healthcare challenges of the future change to meet the requirements of increased life expectancy.

11.1.1 The cell therapy industry: an overview

'Cells as therapies' is the most powerful and disruptive healthcare platform technology to emerge in the twenty-first century (Brindley and Mason, 2012). At present, however, its potential is being constrained by a number of barriers to commercialisation (Plagnol *et al.*, 2009), including a lack of standardised, scalable and cost-effective manufacturing strategies (Mason and Hoare, 2006; Kirouac and Zandstra, 2008; James, 2011).

While the global healthcare sector generates significant societal benefit, it is not an entirely altruistic pursuit; it must also generate revenue for shareholders. Therefore, the ability to manufacture revolutionary cell-based therapeutics alone is insufficient to sustain the nascent cell therapy industry (CTI) (Mason *et al.*, 2011). Instead, the ability and capacity to manufacture cell therapies in a *commercially viable manner, and at scale*, is a mission-critical step in the translation of outstanding laboratory science into therapeutic products capable of benefiting humanity in a competitive and sustainable twenty-first-century industry (Brandenberger *et al.*, 2011). Figure 11.1 highlights the key challenges faced by the cell therapy industry in translating the basic science into robust manufacturable products.

Because of the need to manufacture scalable and commercially viable cell therapies, the industry is currently in a state of transition from the development of basic science towards translation (Cooksey, 2006; Lysaght, 2006; Mason and Manzotti, 2010; McKernan *et al.*, 2010) – specifically, the translation of great science into great products, from 'lab to bedside' (Brindley and Davie, 2009). This transition is undoubtedly exciting; however, it is also frustrating for patients and unnerving for investors, especially in view of the severe financial challenges faced by tissue engineering, the then in vogue emerging industry, at the turn of the century (Lysaght, 2006; Mason, 2007). However, it must be stressed that the CTI is distinct from tissue engineering and regenerative medicine (Mason *et al.*, 2011), and therefore will not only employ different core manufacturing platforms, but also have very different commercial characteristics to its predecessors (Rao, 2011). Due in significant part to increases in the efficiency of commercial translation and thus decreasing costs, the CTI is finally coming of age alongside pharmaceuticals, biologics and medical devices as the fourth and final pillar of healthcare (Mason *et al.*, 2011). However, for it to be successful, development of robust bioprocess steps that can be standardised for scalable and reproducible manufacture will be required.

214 Standardisation in cell and tissue engineering

Process and product development technology ⭐	Business models, reimbursement and COGS
To assess emerging cell processing technologies and forecast their impact on the commercialisation process	*To define economic aspects of manufacturing clinical cell products and tools influencing reimbursement and acceptance as standard of care*
Industry education	Clinical development and new product introduction
To construct educational platforms that drive commercialisation objectives with external societies and industry community	*To address unmet patient needs by connecting industry, academia and global regulatory agencies*

11.1 Key challenges in the cell therapy industry. *Note*: Figure adapted from Deans *et al.* (2010) with the star representing the position of this chapter within the sector.

11.1.2 Defining cell therapies, regenerative medicine and tissue engineering

The terminology used to define the regenerative medicine (regen) sector has been crafted and re-defined of late (Mason *et al.*, 2011). It is particularly important to reiterate that cell therapies, tissue engineering and regenerative medicine are distinct terms:

- **Regenerative medicine:** 'The process of replacing or regenerating human cells, tissues or organs to restore or establish normal function' (Mason and Dunnill, 2008).
- **Tissue engineering:** 'The application of the principles of biology and engineering to the development of functional substitutes for damaged tissue' (Langer and Vacanti, 1993).
- **Cell therapy:** 'The therapeutic application of cells regardless of cell type or clinical indication – a platform technology' (Mason *et al.*, 2011).

Each field utilises its own core competencies and underpinning science for the clinical benefit of the patients they serve. Therefore, it is no surprise that in recent years the usage of these terms has changed (see Plate VII in the colour section between pages 134 and 135). Since, at the turn of the millennium, *Time Magazine* identified it as the 'Number 1 job of the future' (Rawe, 2000), the profile of tissue engineering at an industry level has largely been in decline, driven by the high-profile bankruptcy of

© Woodhead Publishing Limited, 2013

Commercial manufacture of cell therapies 215

Advanced Tissue Sciences (Pangarkar *et al.*, 2010). Conversely, the CTI, led by its first blockbuster, Provenge (Dendreon), continues to grow and offer innovative and complementary products to 'Big Pharma' – addressing its declining R&D outputs (Garnier, 2008) and a number of imminent major patent expirations (Denoon and Vollebregt, 2010).

11.1.3 Current state of the cell therapy industry (CTI)

In 2011 the CTI passed a formidable landmark: one billion dollar annual turnover (Mason *et al.*, 2011); and it has now treated in excess of 100 million patients in the USA alone (Evers, 2009). However, the industry is not yet broadly profitable due in significant part to its lack of manufacturing expertise and capacity (Mason and Dunnill, 2009b; Brandenberger, 2011). Therefore, it is essential that all industry stakeholders work to understand and overcome these manufacturing challenges. Standardisation of protocols will greatly facilitate commercial manufacture, by providing highly defined and, where possible, scalable solutions.

Conventional pharmaceutical manufacturing has been driven by a universal or product-based business model. That is to say, those large volumes of a single core product could be manufactured, thus realising economies of scale, and distributed to patients. This is not necessarily translatable to cell-based therapies. There are a number of implications for manufacturing and the business models that can be sustained.

Autologous vs allogeneic cell therapy: the eternal debate

Allogeneic products are especially attractive from a manufacturing and commercial perspective as they are more akin to conventional, non-cell-based pharmaceuticals in terms of their potential for scale-up. Therefore, a single batch could be used to treat large numbers of patients (Mason and Hoare, 2007). This is significant, as batch size produced for clinical testing will likely be very small and costly, but scalable manufacture to increase batch size of the final approved product will be necessary to ensure that it is commercially viable (Rowley, 2010). In addition, scalability of batch size means that quality assurance and quality control testing and their associated costs can be applied to a single batch, irrespective of the number of doses provided by that batch (Fig. 11.2).

However, from a regulatory and risk management perspective, allogeneic cells are also more challenging due to potential issues of immune rejection and the extensive safety testing required to ensure the risk of disease transmission from the cell source is negligible. In fact, 50% of the costs associated with master cell banks are ascribed to safety testing (Rowley, 2010).

© Woodhead Publishing Limited, 2013

216 Standardisation in cell and tissue engineering

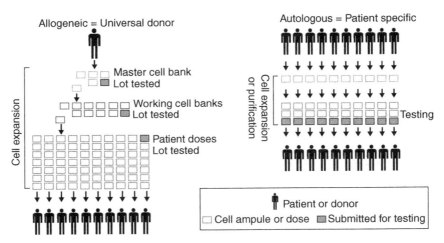

11.2 Testing challenges in cell therapy manufacture (Brandenberger et al., 2011). (This figure is reprinted with permission from *BioProcess International*, March 2011, pp. 30–37.)

Autologous cells offer a number of benefits to patients, including minimised risk of rejection and the potential for personalised medicine. What is more, safety concerns are minimal as the patient to be treated is the source of cell material. However, the service-based business model of autologous therapy is ill-adapted to the established pharmaceutical industry. Even Big Pharma's interpretation of stratified medicines falls well short of bespoke single-patient therapies derived from the patient's own material (Trusheim et al., 2007; Brindley, 2012). Hence, autologous therapies do not offer the same opportunities for significant cost reduction as their allogeneic counterparts, as they pursue a 'scale out' approach that relies on increasing numbers of parallel culture systems, each one designated for a specific patient. Whereas quality control analyses only need to be conducted once for an entire batch of allogeneic cells, for autologous therapy quality control needs to be carried out on every individual batch that is produced. Since each autologous batch will only treat a single patient, the costs associated with quality control analysis and release by a Qualified Person will always be much higher than for a single batch of cell product for allogeneic therapy used to treat many patients (Fig. 11.2). Therefore, it is not uncommon for quality control costs of autologous therapy to account for up to 40% of the selling price.

Differences in scalability and technological requirements for the manufacture of autologous versus allogeneic therapies will continue to fuel debate regarding their respective therapeutic values for many years. A brief overview of the merits and challenges of each type of therapy is provided in Table 11.1 (adapted from an in-depth discussion on the topic by Mason

Commercial manufacture of cell therapies   217

Table 11.1 Potential advantages and challenges of autologous and allogeneic cell therapies

Advantages and benefits	Disadvantages and challenges
Autologous	
• Patient-specific – single patient manufacturing (product of one)	• Variability of source material
• Avoids immune rejection	• Difficult to generate large numbers of cells from either patient-specific somatic or stem cells
• Does not require costly immunosuppression and its associated complications	• Inability to deal with the majority of emergencies
• May be easier to proceed (e.g., no requirement for cell line development)	• Patient throughput will be relatively low
• Reduced start-up costs	• Difficult to address large numbers of patients at reasonable costs
• Avoids embryonic stem cell sources	• Minimal economies of scale
• Simpler regulatory environment	• Biopsy procedure is not without risk to patients
• Avoids non-donor virus and prion transmission concerns	• Any processing failure involves major treatment delays
• Potential for 'point-of-care' processing	
• Could enable independent clinical technology	
• Favoured for bioaesthetic applications	
• Service model-orientated (e.g., embedded in a hospital or clinic)	
• Potentially preferable to patients (self versus non-self debate)	
Allogeneic	
• Producing cells for many patients is more efficient	• Immune rejection may be a major issue
• Potential for scale-up (presently via scale-out)	• Risk of cell abnormalities, particularly with many cycles of *in vitro* replication
• Quality control (QC) can be applied to large lot sizes	• Teratoma (benign tumour) formation risk is a concern (pluripotent cells)
• Existing attachment cell technology for production scale is useful for early clinical trials and orphan low-cell-number therapies	• Provision and consenting of donated cells requires significant time and resources
• Material of high consistency	• Development investment is high
• Allows high patient throughput	• Present lack of manufacturing technology to robustly and cost-effectively commercialise either a blockbuster product or a therapy requiring high cell numbers per application

(*Continued*)

© Woodhead Publishing Limited, 2013

218   Standardisation in cell and tissue engineering

Table 11.1 Continued

Advantages and benefits	Disadvantages and challenges
• Cells are always available • Can address emergency indications • Off-the-shelf availability – 'cells as pills' • Represents a good commercial opportunity for cell suppliers/ contract manufacturing organisations (CMOs) • Requires less clinical time and resources as no patient biopsy needed • Avoids requirement for biopsy consent from severely ill patients • Does not require individual patient biopsy/product segregation, tracking and transport • Commercial product-orientated	

and Dunnill, 2009a). Ultimately, the choice between autologous and allogeneic products is dominated by two decisions. The first addresses which type of cell can produce the greatest therapeutic effect, meeting the required demands of safety, efficacy and purity. The second, crucially for growth and sustainability of the regen sector, addresses which cell type offers the greatest commercial opportunity whilst being culturally and legally acceptable, if both cell types can be used (Brindley, 2012).

Does a protocol need to be standardised for manufacture? In short, yes, due to regulatory requirements. However, the level of standardisation depends very much on the intended therapeutic application. A scalable method for allogeneic therapy that uses automation can, once defined, be 'locked' and then utilised for global manufacture using well-defined, reproducible batch production methods. However, the requirement for autologous 'personalised' medicines utilising relatively small quantities of cells (up to around 10^7 cells per treatment) makes this kind of approach inappropriate.

Standardisation of protocols at the autologous cell manufacture might be limited to ensuring a defined protocol or pathway has been rigorously adhered to in order to meet good manufacturing practice (GMP) standards but with enough leverage for operator judgment to address potential issues relating to biopsy size, cell yield from biopsy, growth capacity and other factors that might significantly impact on yield of product.

Standardisation is also necessary irrespective of the cell source in order to maintain levels of consistency that meet quality control standards across multiple batches or where production is moved from one manufacturing site to another.

© Woodhead Publishing Limited, 2013

Table 11.2 Summary of commercially available cell therapy and regenerative products available as of late 2012

Company	Lead product(s)	Indication	Market launch	Allogeneic/ autologous	Estimated revenue (million $/year)	Current markets
Advanced Biohealing, a Shire company (Dublin, Ireland)	Dermagraft	Diabetic foot ulcers	2001	Allo	>100	US, Canada
Dendreon (Seattle, USA)	Provenge	Prostate cancer	2010	Auto	>100	US
Organogenesis (Canton, USA)	Apligraf	Venous leg ulcers/ diabetic foot ulcers	1998/2000	Allo	>100	US, Saudi Arabia
NuVasive (San Diego, USA)	Osteocel Plus	Skeletal defects	2005	Allo	>50	US
Orthofix (Curaçao, Netherlands, Antilles)	Trinity Evolution	Musculoskeletal defects	2009	Allo	>50	US
Genzyme, a Sanofi company (Boston, USA)	Carticel	Articular cartilage repair	2007	Auto	>10	US, EU
Anterogen (Seoul, South Korea)	Cupistem	Crohn's	2012	Auto	<10	South Korea
Avita Medical (Perth, Australia)	ReCell	Burns, scars	2007	Auto	<10	EU, UK, Australia, Canada
FCB Pharmicell (Seoul, South Korea)	Heartcelligram	Acute myocardial infarction	2011	Auto	<10	South Korea
Genzyme, a Sanofi company (Boston, USA)	Epicel	Severe burns	2007	Allo	<10	US, EU
Japan Tissue Engineering (Gamagori, Japan)	J-TEC Epidermis/ Cartilage/Corneal Epithelium	Burns/cartilage repair/ ocular repair	2012	Auto	<10	Japan
Medipost (Seoul, South Korea)	Caristem	Articular cartilage repair	2012	Allo	<10	South Korea
TiGenix (Leuven, Belgium)	ChondroCelect	Articular cartilage repair	2009	Auto	<10	EU

Note: The number of commercial cell products in 2012 totalled 36, predominantly covering regenerative uses in skin and musculoskeletal tissues. A notable exception is Provenge which is indicated in late-stage prostate cancer.
Source: Reproduced from French *et al.* (2013), Rejuvenation Research, with permission from Mary Ann Liebert Inc.

220   Standardisation in cell and tissue engineering

Current CTI products and those in the pipeline

There are currently 36 approved cell therapies commercially manufactured, as of late 2012 (illustrated in Table 11.2), and a strong pipeline in development. Mason *et al.* (2011) note that even the most conservative estimates suggest that more than one hundred cell therapies will be approved for sale within a decade.

This is a remarkable achievement, firmly underlining the need for investment into this exciting space. Crucially for cell therapy bioprocessing of the future, the development and standardisation of carefully defined process steps that are amenable to scale-up for translation from laboratory to commercial-scale manufacturing is needed to ensure that cost-effective manufacture and rigorous release criteria can be achieved for each batch of cell therapy product.

11.2   The transition from laboratory to commercial-scale manufacture of cell therapies

The vast majority of regenerative medicines, including cell therapies, are still produced at a laboratory scale (Caine, 2011) and this is true not only for autologous therapies but also for allogeneic therapies currently in clinical trials, which are still often produced using flask-based adherent culture methods (Rowley, 2010). Furthermore, these methods rely upon manual processing, which in itself brings a whole raft of potential barriers to scalable, cost-effective manufacture. Manual processing does have a niche and can be highly effective, particularly during small-scale bioprocessing where only relatively small quantities of cells are required (Du Moulin and Morohashi, 2000) or during product development (Ratcliffe *et al.*, 2011) where protocol development or manipulation is required to enhance the product. This is often the case during autologous cell therapy production, where variation in biopsy size and biopsy site (clinician-dependent factors) coupled with variability in number of cells liberated from the biopsy and their growth potential (dependent on biological age, disease status and severity), and variability between manual operators can all collectively affect resultant cell product phenotype (Ratcliffe *et al.*, 2011). Here, whilst the manual operators themselves can introduce variability, the employment of skilled and experienced technicians is extremely valuable when working with 'bespoke' autologous material that does not conform to a 'one-size-fits-all' protocol that is amenable to semi- or fully-automated scalable manufacturing methods. However, small-scale manufacturing techniques for bespoke products is a niche market that in the short-term is envisaged to sustain a small market where therapies are produced in a clinical service environment for small numbers of patients.

© Woodhead Publishing Limited, 2013

Traditional laboratory-scale manufacturing strategies such as flask-based cell culture faces many limitations for the long-term realisation of scalable cell therapy bioprocessing solutions and certainly cannot sustain global markets. Firstly, scalability of the manual processing required to produce larger quantities of cell product (particularly for allogeneic therapy) often relies upon employing increasing numbers of manual operators to achieve the desired scale of production. However, with this comes increasing operator-dependent variation that can impede bioprocess protocol standardisation. Even subtle process variability, such as number of passes through a pipette tip or capillary during manual processing, can impact on cell viability and resultant phenotype (Zoro *et al.*, 2008; Brindley *et al.*, 2011; Mulhall *et al.*, 2011) and therefore be counter-productive to protocol standardisation. Secondly, manual operators are considered the main source of microbial contamination during cell bioprocessing, particularly with respect to the dissemination of airborne infective agents (Reinmuller and Ljungqvist, 2003; Mason and Hoare, 2007) and this can potentially lead to loss of entire batches of product, which is highly costly to a relatively small-scale manufacturing facility.

11.2.1 Considerations for translation from laboratory bench to factory floor

Limited manufacturing capacity in emerging healthcare technologies is not unusual. For example, the blockbuster Enbrel (Etanercept), developed by Immunex (Seattle, WA, USA, subsequently acquired by Amgen, Thousand Oaks, CA), single-handedly depleted all manufacturing capacity in the early monoclonal antibody (mAb) field (Thiel, 2004). In hindsight, this was unsurprising as mAbs were an untested technology and Enbrel was the first such product approved for treatment of rheumatoid arthritis. However, it did make investors question the feasibility of the broader, and now very successful, biologics industry, and cost its manufacturer, Amgen, billions of dollars in lost revenue and valuable time to exploit the marketing exclusivity conferred by finite patents (Kamarck, 2006).

While the CTI today shares many of the 'growing pains' of emerging industries, including mAbs, it also poses a number of its own unique challenges, due primarily to the greatly increased complexity of the living cell product compared to expressed protein products. From a manufacturing perspective issues relating to intellectual property strategies, the influence of 'legacy products', the need for process robustness and the different requirements for autologous *vs* allogeneic product manufacture all pose a challenge for cell therapy production.

222   Standardisation in cell and tissue engineering

Intellectual property (IP)

mAbs, like their predecessor, small molecules, have established and robust IP strategies. This is, at least in part, because they were entirely manufactured using genetic engineering techniques and without doubt are unlikely to occur in nature. Therefore, innovative mAbs satisfied three key requirements for patent protection: (1) no prior art, (2) not obvious to a person skilled in the art and (3) not naturally occurring. However, cell therapies have proved somewhat more challenging.

Whilst it may in some circumstances be possible, it is overall undoubtedly more difficult to patent a cell line or a key characteristic of one, than it is to patent a genetically engineered mAb or chemically synthesised small molecule (Rios, 2011). This is due, most notably, to the 'naturally occurring' nature of cells and the serious ethical questions raised by stem cell science. Moreover, in the case of point-of-care bioprocessing systems, where little or no cellular manipulation occurs in the production of the final therapeutic, except maybe pre-conditioning with a recombinant growth factor to enhance a desired cell phenotype, it is only possible to obtain IP protection on the processing equipment itself and not the therapy produced. It is also important to note that it is not possible to patent a medical procedure. Therefore, on the whole, intellectual property in the CTI is dominated by trade secrets, accompanied by the mantra 'the product is the process' (Mason and Hoare, 2007). This industry feature is challenging to researchers in the space as very little literature is published concerning the large-scale manufacture of cell therapies, thus perpetuating the void between CTI-focused academia and industry. For example, the first CTI monograph for Provenge (Dendreon) has only recently been published (VA-PBM-Services, 2011).

Legacy products

A number of early CTI products, especially autologous products, attracted media coverage that was disproportionate to their degree of commercial success. Whilst some of the first-generation products were not commercially feasible due to their poor scalability and highly labour-intensive manufacturing processes, they did provide a very valuable introduction to the challenges and bottlenecks that the CTI needs to address in order to produce scalable processing methods.

Process robustness and clinical success

Conventionally, for therapeutic technologies, manufacturing efficiency and process robustness have had the greatest impact on production capacity and profitability. However, in the case of the cell therapy industry, where there is greater bioprocess and product variability between batches due to statistical

Commercial manufacture of cell therapies 223

variation amongst the multiple key input parameters, there is a fundamental link between process robustness and clinical success (Burger, 2003; Brandenberger, 2011). Standardising protocols and then firmly adhering to them across multiple batch production during cell therapy manufacture will help to reduce the product variability and maintain consistent levels of clinical success. Even where a protocol is seemingly adhered to, subtle operator-mediated differences in manual handling procedures, such as number of pipette/capillary passes, rate of pipetting, processing time during passaging and its impact on CO_2 levels and therefore pH can impact on the phenotype of the cell product (Veraitch *et al.*, 2008; Zoro *et al.*, 2008; Liu *et al.*, 2010; Brindley *et al.*, 2011; Mulhall *et al.*, 2011). Therefore, it is vital from a regulatory and commercial perspective that robust and commercially viable manufacturing processes are developed early in clinical studies before processes becomes 'locked down' by the regulators, and costly comparability studies are required before the product is permitted to continue towards market (Mason and Manzotti, 2010). Any process changes introduced during pre-market approval of a cell therapy candidate would require re-validation to confirm that the safety and efficacy of the product have not been compromised, thus further increasing costs and delaying market release (Mason and Hoare, 2007; Rowley, 2010; Ratcliffe *et al.*, 2011). In short, it is essential that process development is addressed early in clinical studies in order to save time and money later (Sweet, 2010).

The development rates of autologous vs allogeneic products

From a commercial perspective, the development rates of autologous and allogeneic cell products are very interesting, reflecting differences not only in cell product process requirements and safety but also in commercial investment. Due in part to the CTI's roots in bone marrow transplantation and blood processing, commercial-scale autologous cell therapy processing has been refined at a faster rate than allogeneic processing of at least some types of cell therapy intervention. This is because for autologous cell therapy (such as during autologous chondrocyte transplantation), the bioprocess tools are essentially the same as those used for lab-scale cell culture, albeit in a scaled-out manner and to GMP guidelines. Conversely, the development rate of large-scale allogeneic manufacturing technologies, including suspension cultures, has been slower, due to the major technological step change required in the way adherent human cells for therapy are cultured and the lack of translatability and applicability of equipment and technologies from other fields. To date, attempts at translating technologies from other industries have really been limited to stirred tank bioreactors utilising microcarriers to provide substrate support for attachment of adherent cells (King and Miller, 2007; Brindley *et al.*, 2011; Yeatts and Fisher, 2011).

© Woodhead Publishing Limited, 2013

224　Standardisation in cell and tissue engineering

However, these processes are not always appropriate for scalable production of cell-based therapeutics due to difficultly in liberating the cells from their microcarriers after expansion and reduced cell viability or phenotype modification that results from exposure to flow-induced mechanical stress (Brindley *et al.*, 2011).

11.2.2　Process development

A manufacturing process designed to produce a cell-based therapeutic needs to be optimised prior to standardisation. Process development is essential in order to ensure critical quality attributes of the cell product, namely identity, potency, purity and safety, are optimal (Brandenberger, 2011). The bioprocess environment for human cells contains multiple signals that can impact on cells. To determine the optimal protocol for manufacture of a given cell therapy, the nature of the starting cell material, along with all the chemical, physical and mechanical cues within the bioprocess environment and the timing, sequence and duration of application of those cues will affect cells in different ways either individually, collectively or synergistically (Kirouac and Zandstra, 2008). Simple process development whereby one variable is examined at a time is not appropriate. Instead multifactorial experimental approaches need to be employed to address the huge potential variability in the exact product phenotype (Thomas *et al.*, 2008). One approach taken to meet this requirement is ultra scale-down (USD), whereby bioprocessing is modelled using bench-scale or even micron-scale tools for screening multifactorial parameters to identify optimal process conditions (Willoughby *et al.*, 2004). USD methods have been developed for the analysis of bioprocess parameters for cell therapy production (Markusen *et al.*, 2006; Titmarsh and Cooper-White, 2009; Ratcliffe *et al.*, 2011). These methods are very promising for precisely defining optimal conditions for cell therapy manufacture, such as for expansion and differentiation of human pluripotent stem cells into specific neuronal populations. Even for autologous cell culture, the product characteristics obtained from a single common bioprocess applied across multiple parallel cultures from different donors will facilitate identification of critical operating parameters that will give the best yield and quality of cell product, accounting for the large variability in the start material. Process development for autologous cell therapy processing by manual operators resulting from subtle improvements and optimisation of protocols helps to meet validation requirements. For example, one prominent paper reported that cell growth rates for expansion of autologous chondrocyte procedures were gradually improved from 0.251 PD to 0.311 PD over the first 5 months of manufacture, as procedures were improved and validated (Mayhew *et al.*, 1998).

© Woodhead Publishing Limited, 2013

11.3 Key regulatory requirements for commercial manufacture of cell therapies

The translation of good basic science into a therapeutic product requires, in addition to appropriate bioprocess tools for industrial manufacture, many regulatory requirements to be met. The regulatory requirements for cell therapy manufacture are extensive, reflecting the complexity of the product. The key concerns are associated with maintaining high consistency and quality of the product and eliminating factors that could compromise safety. The summary below is designed to give a broad overview of the main regulatory concerns.

11.3.1 Reproducibility and comparability

The mantra 'the product is the process' (Mason and Hoare, 2006, 2007) is undoubtedly a fundamental concept in cell therapy manufacture, and the intrinsic link between the bioprocess environment, product quality and regulatory compliance cannot be stressed enough. Moreover, it is essential to ensure 'product consistency, quality, and purity by ensuring that the manufacturing process remains substantially the same over time' (Greenwood, 2010). Protocols need to be standardised as much as possible so that inherent statistical variation between operating variables is minimised and both the process and product are robust and reproducible. Process robustness is an essential consideration for ensuring commercial viability of products within the emerging cell therapy industry (Carmen *et al.*, 2011). Manual protocols are very difficult to standardise due to multiple process steps being susceptible to operator variation. Automation represents a potential solution to the challenges and barriers to cell therapy commercialisation with respect to standardising the bioprocess environment and operator functions to create reproducible processes that are comparable between batches and lend themselves to up-scaling. By eliminating operator variability using highly precise programmed robotics, one of the main sources of statistical variation is significantly reduced, if not totally eliminated.

11.3.2 Serum elimination

It is highly desirable to create cell therapy products that are cultured in conditions free from animal products, mainly to reduce safety concerns associated with xenoviral transfer and contamination. This also includes removal of foetal calf serum (FCS) from culture medium. There is a misconception, however, that regulatory requirements stipulate that cell therapeutics are

226 Standardisation in cell and tissue engineering

processed in serum-free conditions (Ratcliffe *et al.*, 2011). Current regulatory guidelines allow the use of serum from traceable animals from specific geographic locations to be prepared under strict manufacturing guidelines in place to reduce risk of transmission of transmissible spongiform encephalopathies (EAEMP, 2004), so while the costs are higher than using standard FCS, the translation from a laboratory method to one that is acceptable for manufacture of cell therapy products is reasonably straightforward compared to use of synthetic serum replacements. One of the downsides is that the composition of FCS is still not completely known and there can be considerable batch-to-batch variation, so in terms of scalable and robust bioprocessing, FCS is yet one more input variable that will impact on process output (Ratcliffe *et al.*, 2011). In order to create standardised protocols that yield a cell product reproducibly, time after time and across multiple different facilities, use of serum replacement methods that are free from animal products will be much more favourable.

11.3.3 Extended culture duration

Two prominent examples of disease targets requiring scalable cell therapy manufacture include type I diabetes and myocardial infarction. It is predicted that for successful treatment of a single patient with human embryonic stem cell-derived pancreatic β-cells, a dose of 10^8 cells will be required (Mimeault and Batra, 2006) and to deliver functional cardiomyocytes to the injured myocardium could require a staggering 10^9 cells (McDevitt and Palecek, 2008). Achieving this level of scale, taking into account the low level of efficiency of differentiation purification protocols, is a daunting task. Using adult mesenchymal stem cells (MSCs) to achieve this level of scale is simply not possible. MSCs from young donors normally only undergo approximately 40 population doublings (PD) before entering a state of replicative senescence, reducing to around 25 PD in the aged (Stenderup *et al.*, 2003). This is predominantly due to progressive attrition of telomeres at each cell division event (Simonsen *et al.*, 2002). Lifespan extension by re-activation of the telomerase promoter has been shown to immortalise MSCs, something that would potentially be of interest for achieving scalable methods of therapeutic cell expansion. However, karyotypic instability after 250 PD has been reported (Serakinci *et al.*, 2006), echoing the concerns that long-term pluripotent cell culture has raised. In the case of embryonic stem cells or induced pluripotent stem cells, the challenges are even greater as karyotypic stability is very hard to maintain. Therefore, the advantage of high cell number achievable from pluripotent cells in comparison to adult somatic cells is negated by concerns of karyotypic instability that often results from long-term cell expansion.

© Woodhead Publishing Limited, 2013

11.3.4 Product safety

Whilst living cell-based therapies offer considerable promise as a health-care platform for addressing currently unmet clinical needs, they also raise greater concern regarding product safety than small molecules or inert biologics primarily because the product is 'live' and therefore has the potential to change in response to its environment. The US Food and Drug Administration (FDA) states that 'in-process testing as well as testing of final formulated product (prior to cryopreservation) and of thawed products is required in order to ensure the quality and consistency of the product. The testing shall include sterility (bacteria), mycoplasma, as well as testing for adventitious viral agents' (FDA, 2008). This can be challenging where a product has a limited lifespan – for example, Provenge (Dendreon) is administered to the patient prior to the results of sterility tests (Higano *et al.*, 2010). Therefore, patients have to give informed consent for treatment, in case the product is contaminated upon delivery. Most conventional microbiological testing methods are inadequate where cell therapy has a short shelf life (Rayment and Williams, 2010). However, the use of closed systems can be advantageous here, as there are fewer opportunities for contamination to occur when processing is carried out in a closed environment with minimal operator intervention.

A second form of cell product safety testing, particularly in relation to allogeneic therapy, is tumorigenic capacity of the cells. A cell therapy derived from a pluripotent source may still have a proportion of undifferentiated cells within the population (Fong *et al.*, 2010). Transition towards clinic can only be made when the pluripotent cell-derived product has been proven safe. Teratoma formation studies using the cell product in immunocompromised animals are frequently performed. In studies where no teratoma formation has been observed more than one month after transplantation, it has been hypothesised where pluripotent cells are differentiated *in vitro* for long durations, the risk of teratoma formation is lower as the chance of pluripotent cells remaining is lower (Brederlau *et al.*, 2006; Laflamme *et al.*, 2007). In addition to prolonged differentiation times, protocols aiming to ensure safety of pluripotent stem cell-derived cell therapy candidates have included selective antibodies carrying killer molecules or apoptosis agents that specifically target contaminating pluripotent cells, cell sorting techniques and density gradient centrifugation (Fong *et al.*, 2010).

11.3.5 Process versus product safety

An important distinction in cell therapy manufacturing arises in the area of safety. There are two main concerns: process safety and product safety.

228 Standardisation in cell and tissue engineering

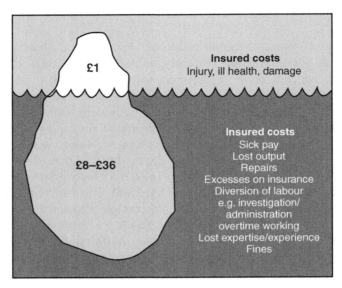

11.3 The 'Cost Iceberg' (University of Nottingham, Safety Department).

While to some extent inextricably linked, they are distinct and both worthy of serious consideration. Robust health and safety provision is important in any organisation in order to protect its employees, equipment, reputation and finances – as illustrated by the 'Cost Iceberg' (see Fig. 11.3). Moreover, in bioprocessing of pharmaceuticals it is important to meet a manufacturer's moral and legal obligations beyond the factory gates, popularised as a 'cradle-to-grave' approach (Mahgerefteh, 2009). For example, biopharmaceutical manufacturers are responsible for product safety throughout distribution to the clinician, and following administration, its long-term pharmacokinetic effects *in vivo*. In addition to this, it is not unusual for safety audits for new manufacturing facilities alone to consist of tens of thousands of pages.

11.3.6 Essential steps towards a commercial process

A number of essential steps have to be taken in order to translate a good lab-scale cell therapy production protocol into a commercially viable process that is conducted according to GMP guidelines. A summary of some of the necessary steps, along with an overview of the reasons for each step, is provided in Table 11.3.

Commercial manufacture of cell therapies   229

Table 11.3 Essential steps needed in the translation and commercialisation of cell therapy

Essential steps towards a commercial process	Reason
1 Remove serum from process	• High batch variation impacts on standardisation of protocols and reproducibility • Risk of transmittable disease
2 Utilise closed systems and equipment wherever possible	• Reduce risk of microbial contamination or cross-contamination between samples
3 Remove antibiotics from process	• Prevent masking of underlying low-level contamination and demonstrate a 'clean' process
4 Focus on larger expansion technologies than t-flasks e.g. 40 layer cell factories and HyperFlasks – especially for allogeneic products	• Necessary to achieve the level of expansion needed to treat multiple patients from a single batch or where single-patient doses needed are high ($\sim10^8$–10^9 cells per treatment)
5 Eliminate manual processing wherever possible	• Reduce operator-dependent variability inherent to manual processing • Reduce contamination risk
6 Conduct process(es) in the minimum background that is regulatorily-permitted and preserves critical quality attributes (CQAs)	• Simplicity will facilitate easier standardisation of protocols and consistency in cell therapy product
7 Focus on achieving process standardisation, particularly in terms of media and reagents AND product personalisation for patients	• Standardisation should lead to easier optimisation of processes and production of cells and facilitate troubleshooting where problems relating to the processing arise
8 Fully characterise all cell lines (James, 2011)	• To create a better understanding of the cellular material and facilitate regulatory assessment
9 Assay development	• To validate the process and provide quality control measures for the cell product in terms of safety and identity
10 Consider supply chain issues including traceability and shelf life	• To determine the commercial feasibility of the therapy as a product
11 Clinician and pharmacist training – especially for autologous products where a high-quality biopsy is required	• To ensure the best chances of therapy success by optimising crucial stages of the 'whole bioprocess' that are conducted outside of the laboratory
12 Establish reference standards	• As a 'standard' for measuring batches of the cell therapy product against to ensure quality

(Continued)

© Woodhead Publishing Limited, 2013

230 Standardisation in cell and tissue engineering

Table 11.3 Continued

Essential steps towards a commercial process	Reason
13 Maximise facility and resource utilisation (James, 2011)	• To enhance commercial viability of the product by ensuring efficiency within the manufacturing process
14 Provide appropriate personnel training, including QP	• To have specialist understanding of regulatory framework and requirements for cell therapy product release
15 Focus on cost of goods sold and reimbursement structures early (Maziarz and Driscoll, 2011; Rowley, 2010b)	• To ensure commercial success, a detailed understanding of the costs and reimbursement structures is needed early to determine financial viability of a cell therapy

11.4 Cell-based therapy versus monoclonal antibody therapies: lessons from existing biopharmaceutical manufacture

A number of comparisons, including commercial and manufacturing, have been made between the emerging cell therapy industry and the established mAb industry. It is a recurring theme throughout academia and industry and even though there are many highly significant differences between them, there are valuable lessons that can be learned. Therefore, an overview of fundamental manufacturing similarities and differences is included in Table 11.4.

11.4.1 The emergence of a cell therapy platform manufacturing process

There is a clear tendency in the development cycle of manufacturing industries for the production of 'second-generation products' that adopt an industry-wide platform process. This is particularly true in the manufacture of high-volume and high unit-value products such as pharmaceuticals (Nelson *et al.*, 2010). Early indications suggest that the manufacture of cellular therapies will also follow a similar trend; but is this necessarily advantageous or simply convenient? An emerging cell therapy platform process is presented in Fig. 11.4. The exact protocol required to produce a cell therapy will depend upon the specific cell product that is needed and the source of the starting cell material, but the sequence of defined process steps required to derive the product will generally be consistent.

© Woodhead Publishing Limited, 2013

Commercial manufacture of cell therapies   231

Table 11.4 Current overview of mAb versus cell therapy manufacturing processes

	mAb	Cell therapy
Start and end points	*Start point:* cells from cell bank *End point:* vialling	*Start point:* Allogeneic: master cell bank Autologous: biopsy/aspirate Other: blood sample, etc. *End point:* Live cell products: injection/implantation into patient Cryopreserved: vialling
Product	*Dose:* high dosage over a long time period *Product nature:* 'inert' *Heterogeneity:* variety of glycoforms present	*Dose:* typically only one treatment or application *Product nature:* live, dynamic and responsive to applied stimuli *Heterogeneity:* difficult to characterise impact of different cell types in final population
Single-use components	Single-use components for seed stages; but stainless steel at large scale	Small-scale processes – therefore single-use components are dominant
Process	*Process automation:* largely automated *Platform process:* Generic process dominates *Process delays:* stable product – minimal problems	*Process automation:* partial automation – labour-intensive *Platform process:* variety of processes in use *Process delays:* sensitive, dynamic product – therefore delays are a serious problem
Key sources of variability	Product titre varies with cell line	Initial cell sources, e.g. biopsy, varies in terms of cell number and viability
Cell media and culture	Defined, animal-free media	Complex media containing supplements, e.g. amino acids, growth factors and ligands 'Legacy products' are not animal-free
Recovery and purification	Dominated by chromatography operations	Expanded cells: volume reduction Differentiated cells: cell sorting but limited scope for scale-up
Transportation and storage	Usually ambient or refrigerated and stable for several months	Cryopreserved or live: therefore, limited shelf life and need for high-grade courier service

Acknowledgement: The authors wish to extend their thanks to S. S. Farid, C. Mason and N. Davie of UCL ACBE and D. Smith, P. Vanek and K. Warren of Lonza, Walkersville, MA, USA for their long-term contributions towards the production of this table.

© Woodhead Publishing Limited, 2013

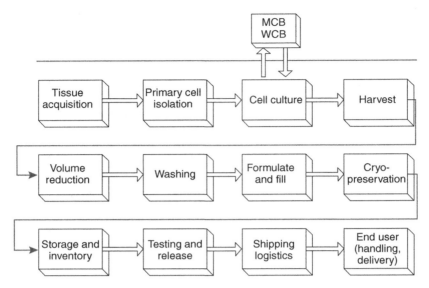

11.4 An emerging cell therapy platform process (Brandenberger *et al.*, 2011). Reprinted with permission from *BioProcess International*, March 2011, pp. 30–37.

Several key factors tend to influence the utilisation of platform processes for manufacture.

Cost

All pharmaceutical manufacture for the healthcare sector is driven by a desire to meet the needs of the patient and the shareholder. Unsurprisingly, therefore, there are potential cost benefits to be realised from the adoption of generic manufacturing processes. For example, by utilising existing manufacturing infrastructure the 'time to market' for new products can be reduced; and routine clinical practice and reimbursement can be achieved more rapidly, which is critical for a product to be commercially successful. Moreover, platform technologies reduce the need for extensive initial investment in manufacturing facilities for biotech start-ups. These scarce funds can be conserved to create value by developing product pipelines and tailoring existing products for the treatment of new indications. This has contributed to the dominance of the VIPCO business model in the cell therapy industry and has provided opportunities for contract manufacturing organisations (CMOs) to generate income through the sharing of expertise and facilities to create multiple different therapies.

Process

It is evident that a failure to effectively manage and communicate risks, from manufacturing to exit, has deterred investors from substantially supporting the emerging cell therapy industry to date (Giebel, 2005). The application of platform processes, which proved very successful in the mAb manufacturing industry, provides an opportunity to manage manufacturing risk by, for example, improving process reliability and reducing the likelihood and impact of processing delays. Furthermore, platform processes that have proven reliable for the manufacture of one product type are viewed favourably as a platform for manufacture of other similar products, as 'proof of principle' is already demonstrated. A modular process also confers process flexibility, in particular with respect to the ease of optimisation of the process and characterisation of the resulting cell product (Carson, 2005).

Practical and logistical advantages

The practical advantages of platform technologies are numerous. For example, all personnel within an organisation become familiar with the prevailing platform at all stages of development. Therefore, production strategies can be tailored to product requirements early in lab-based studies as this informs multifactorial design of experimentation (DoE) assessment and provides a planning tool for pipeline products (Sommerfield, 2005). The need for frequent staff training is also reduced, increasing employee productivity. A further advantage is that generic processes are not site-dependent. Therefore, production can be transferred to locations where the cost of specialist labour is lowest. This is also of assistance to CMOs in planning to meet industry demands in the future, thereby preventing manufacturing shortages such as those currently experienced by Dendreon in the production of their Provenge therapy (Glasier, 2011).

The negatives

As with most bioprocessing activities, any form of standardisation is accompanied by some form of trade-off. Clearly the optimal processing demands of one biological entity will differ from those of another. Existing biopharmaceuticals, for example monoclonal antibodies, have been genetically re-engineered and selectively screened for their efficacy and amenability to industrial bioprocessing. Conversely, in the case of cell therapy manufacture, the starting cellular material (particularly for an autologous product) is selected on the basis of *in vivo* function and donor site availability as opposed to any potentially advantageous bioprocessing characteristics. Therefore, there is a fundamental need for flexibility in production strategies to accommodate the intrinsic biochemical variability, or heterogeneity of the

234   Standardisation in cell and tissue engineering

starting cell population. However, in many instances platform flexibility will be insufficient to accommodate the demands of particular cell types, leading to poor process efficiency or a potentially efficacious cell therapy candidate being abandoned due to the potential cost of producing a bespoke facility. This is particularly true in the case of many biotech start-up companies who simply do not have the financial resources to construct bespoke facilities. In this respect, standardisation can to some degree suppress the innovation required to take cell-based therapies towards routine clinical practice.

Given the short shelf life of some cell therapy products, such as Provenge (Dendreon) that only has an eighteen-hour shelf life, large outsourced industrial facilities may not always be suitable manufacturing centres. In the case of products with very short shelf lives, they will have to be manufactured in smaller regional centres 'just in time' for administration, indicating a need for small-scale facilities to be integrated into existing hospital GMP environments, something that is already underway in a number of clinical centres across Europe.

11.4.2 Clinical trial structures and manufacturing demand

A key driver of manufacturing development is the product demand required in order to facilitate clinical trials, as this can be a key and expensive bottleneck towards market. Clinical trial manufacturing demands are similar between small molecules and mAbs; however, the demands of cell therapies are not as great, with far fewer patients recruited for clinical trial phases (Table 11.5). The manufacturing demand for cell therapies can be as little as a tenth of that of small molecules. This is principally because of the unique aspects of cell therapy clinical trials, including the fact that 'first in man' phase one trials are not conducted in healthy volunteers, rather in existing sufferers.

Table 11.5 Clinical trial patient numbers: cell-based versus conventional therapies

Clinical trial stage	Typical patient numbers (based on a transient cell-based autologous therapy*)	Conventional clinical trial patient numbers (e.g., mAbs)
Phase I	8–10	20–100
Phase IIa	20–25	100–350
Phase IIb	25–30	200–700
Phase III	30–40	300–3000

*Permanent cell-based therapies and allogeneic therapies will generally require a greater number of patients but still significantly less than for a typical pharmaceutical candidate molecule.
Source: Brindley and Mason, 2012.

© Woodhead Publishing Limited, 2013

11.5 Conclusion

Clearly, many factors have to be considered when translating basic cell biology research to an industrial manufacturing process that conforms to regulatory requirements. Unlike small molecules or 'inert' biologics such as mAbs, the cell product is live and may be altered by conditions it experiences during bioprocessing. Many conditions within the bioprocess environment are carefully monitored and controlled to ensure minimal impact and standardisation of the manufacturing protocol to ensure consistency amongst batches. Lessons can be learned from other industries, such as mAb manufacturing, in terms of platform technology adoption and preparation of the final product. However, the caveat for cell therapy being that the cell is a living product, reminds us that the cell therapy industry is unique and as such, requires a fresh approach to protocol development for commercial manufacturing.

There is still much more work to be done in order to achieve successful translation of good laboratory science underpinning cell therapy candidates to routine clinical practice. However, the field (industrialisation as well as basic science) is progressing rapidly, leaving no doubt that the emerging cell therapy industry will be a fundamental pillar of healthcare for the future.

11.6 References

Brandenberger, R., Burger, S., Campbell, A., Fong, F., Lapinska, E. and Rowley, J.A. 2011. Cell therapy bioprocessing. *BioProcess International*, Supplement, **9**, 30–37.

Brederlau, A., Correia, A. S., Anisimov, S. V., Elmi, M., Paul, G., Roybon, L., Morizane, A., Bergquist, F., Riebe, I., Nannmark, U., Carta, M., Hanse, E., Takahashi, J., Sasai, Y., Funa, K., Brundin, P., Eriksson, P. S. and Li, J. Y. 2006. Transplantation of human embryonic stem cell-derived cells to a rat model of Parkinson's disease: effect of in vitro differentiation on graft survival and teratoma formation. *Stem Cells*, **24**, 1433–1440.

Brindley, D. and Davie, N. 2009. Regenerative medicine through a crisis: social perception and the financial reality. *Rejuvenation Research*, **12**, 455–461.

Brindley, D., Moorthy, K., Lee, J. H., Mason, C., Kim, H. W. and Wall, I. 2011. Bioprocess forces and their impact on cell behavior: implications for bone regeneration therapy. *Journal of Tissue Engineering*, **2011**, 620247.

Brindley, D. A. and Mason, C. 2012. Cell therapy commercialisation. In: Atala, A. (ed.) *Progenitor and Stem Cell Technologies and Therapies*. Cambridge, UK: Woodhead Publishing Limited.

Burger, S. R. 2003. Current regulatory issues in cell and tissue therapy. *Cytotherapy*, **5**, 289–298.

Caine, B. and Montgomerry, S.A. 2011. Building a bridge to commercial success. *Bioprocess International*, **9**(S1), 3.

Carmen, J., Burger, S.R., Mccaman, M. and Rowley, J.A. 2012. Developing assays to address identity, poteency, purity and safety: cell characterization in cell therapy process development. *Regenerative Medicine*, **7**, 85–100.

236 Standardisation in cell and tissue engineering

Carson, K. L. 2005. Flexibility–the guiding principle for antibody manufacturing. *Nature Biotechnology*, **23**, 1054–1058.

Cooksey, D. 2006. A review of UK health research funding. London, UK: Her Majesty's Stationary Office.

Deans, R., Gunter, K. C., Allsopp, T., Bonyhadi, M., Burger, S. R., Carpenter, M., Clark, T., Cox, C. S., Driscoll, D., Field, E., Huss, R., Lardenoije, R., Lodie, T. A., Mason, C., Neubiser, R., Rasko, J. E., Rowley, J. and Maziarz, R. T. 2010. A changing time: the International Society for Cellular Therapy embraces its industry members. *Cytotherapy*, **12**, 853–856.

Denoon, A. and Vollebregt, E. 2010. Can regenerative medicine save Big Pharma's business model from the patent cliff? *Regenerative Medicine*, **5**, 687–690.

Du Moulin, G. and Morohashi, M. 2000. Development of a regulatory strategy for the cellular therapies: an American perspective. *Materials Science and Engineering: C-Biomimetic and Supramolecular Systems*, **13**, 15–17.

EAEMP. 2004. *Note for Guidance on Minimising Risk of Transmitting Animal Spongiform Encephalopathy Agents via Human and Veterinary Medicinal Products*. London: European Agency for the Evaluation of Medicinal Products.

Evers, P. 2009. *Advances in the Stem Cell Industry*. Global Business Insights. http://www.ebook3000.com/Business-Insights---Advances-in-the-Stem-Cell-Industry_134613.html

FDA 2008. *Guidance for FDA Reviewers and Sponsors: Content and Review of Chemistry, Manufacturing and Control (CMC). Information for Human Somatic Cell Therapy Investigational New Drug Applications (INDs)*. Rockville, USA: United States Food and Drug Administration.

Fong, C. Y., Gauthaman, K. and Bongso, A. 2010. Teratomas from pluripotent stem cells: a clinical hurdle. *Journal of Cellular Biochemistry*, **111**, 769–781.

Garnier, J. P. 2008. Rebuilding the R&D engine in big pharma. *Harvard Business Review*, **86**, 68–70, 72–76, 128.

Giebel, L. B. 2005. Stem cells–a hard sell to investors. *Nature Biotechnology*, **23**, 798–800.

Glasier, V. 2011. status of cell and gene therapy keeps vacillating. *Genetic Engineering and Biotechnology News*, **11**, 1–3.

Greenwood, J. 2010. *How do Drugs and Biologics Differ?* (Online). Available: http://www.bio.org/node/53.

Higano, C. S., Small, E. J., Schellhammer, P., Yasothan, U., Gubernick, S., Kirkpatrick, P. and Kantoff, P. W. 2010. Sipuleucel-T. *Nature Reviews. Drug Discovery*, **9**, 513–514.

James, D. 2011. Therapies of tomorrow require more than factories from the past. *Bioprocess International*, **9**(Supplement 1), 4–12.

Kamarck, M. E. 2006. Building biomanufacturing capacity–the chapter and verse. *Nature Biotechnology*, **24**, 503–505.

King, J. A. and Miller, W. M. 2007. Bioreactor development for stem cell expansion and controlled differentiation. *Current Opinion in Chemical Biology*, **11**, 394–398.

Kirouac, D. C. and Zandstra, P. W. 2008. The systematic production of cells for cell therapies. *Cell Stem Cell*, **3**, 369–381.

Laflamme, M. A., Chen, K. Y., Naumova, A. V., Muskheli, V., Fugate, J. A., Dupras, S. K., Reinecke, H., Xu, C., Hassanipour, M., Police, S., O'sullivan, C., Collins, L., Chen, Y., Minami, E., Gill, E. A., Ueno, S., Yuan, C., Gold, J. and Murry, C. E. 2007.

© Woodhead Publishing Limited, 2013

Cardiomyocytes derived from human embryonic stem cells in pro-survival factors enhance function of infarcted rat hearts. *Nature Biotechnology*, **25**, 1015–1024.

Langer, R. and Vacanti, J. P. 1993. Tissue engineering. *Science*, **260**, 920–926.

Liu, Y., Hourd, P., Chandra, A. and Williams, D. J. 2010. Human cell culture process capability: a comparison of manual and automated production. *Journal of Tissue Engineering and Regenerative Medicine*, **4**, 45–54.

Lysaght, M. J. 2006. *Tissue Engineering: Great Expectation*. London Regenerative Medicine Network Event, London, UK.

Mahgerefteh, H. A. B., S. 2009. From cradle to grave: High pressure phase equilibrium behaviour of CO_2 during CCS. *Thermodynamic and Transport Properties Under Pressure Session*. AIChemE, New York, NY, USA.

Markusen, J. F., Mason, C., Hull, D. A., Town, M. A., Tabor, A. B., Clements, M., Boshoff, C. H. & Dunnill, P. 2006. Behavior of adult human mesenchymal stem cells entrapped in alginate-GRGDY beads. *Tissue Engineering*, **12**, 821–830.

Mason, C. 2007. Regenerative medicine 2.0. *Regenerative Medicine*, **2**, 11–18.

Mason, C., Brindley, D. A., Culme-Seymour, E. J. and Davie, N. L. 2011. Cell therapy industry: billion dollar global business with unlimited potential. *Regenerative Medicine*, **6**, 265–272.

Mason, C. and Dunnill, P. 2008. A brief definition of regenerative medicine. *Regenerative Medicine*, **3**, 1–5.

Mason, C. and Dunnill, P. 2009a. Assessing the value of autologous and allogeneic cells for regenerative medicine. *Regenerative Medicine*, **4**, 835–853.

Mason, C. and Dunnill, P. 2009b. Quantities of cells used for regenerative medicine and some implications for clinicians and bioprocessors. *Regenerative Medicine*, **4**, 153–157.

Mason, C. and Hoare, M. 2006. Regenerative medicine bioprocessing: the need to learn from the experience of other fields. *Regenerative Medicine*, **1**, 615–23.

Mason, C. and Hoare, M. 2007. Regenerative medicine bioprocessing: building a conceptual framework based on early studies. *Tissue Engineering*, **13**, 301–311.

Mason, C. and Manzotti, E. 2010. The Translation Cycle: round and round in cycles is the only way forward for regenerative medicine. *Regenerative Medicine*, **5**, 153–155.

Mayhew, T. A., Williams, G. R., Senica, M. A., Kuniholm, G. and Du Moulin, G. C. 1998. Validation of a quality assurance program for autologous cultured chondrocyte implantation. *Tissue Engineering*, **4**, 325–334.

Maziarz, R. T. and Driscoll, D. 2011. Hematopoietic stem cell transplantation and implications for cell therapy reimbursement. *Cell Stem Cell*, **8**, 609–612.

McDevitt, T. C. and Palecek, S. P. 2008. Innovation in the culture and derivation of pluripotent human stem cells. *Current Opinion in Biotechnology*, **19**, 527–533.

McKernan, R., McNeish, J. and Smith, D. 2010. Pharma's developing interest in stem cells. *Cell Stem Cell*, **6**, 517–520.

Mimeault, M. and Batra, S. K. 2006. Concise review: recent advances on the significance of stem cells in tissue regeneration and cancer therapies. *Stem Cells*, **24**, 2319–2345.

Mulhall, H., Patel, M., Alqahtani, K., Mason, C., Lewis, M. P. and Wall, I. 2011. Effect of capillary shear stress on recovery and osteogenic differentiation of muscle-derived precursor cell populations. *Journal of Tissue Engineering and Regenerative Medicine*, **5**, 629–635.

238 Standardisation in cell and tissue engineering

Nelson, A. L., Dhimolea, E. and Reichert, J. M. 2010. Development trends for human monoclonal antibody therapeutics. *Nature reviews. Drug Discovery*, **9**, 767–774.

Pangarkar, N., Pharoah, M., Nigam, A., Hutmacher, D. W. and Champ, S. 2010. Advanced Tissue Sciences Inc.: learning from the past, a case study for regenerative medicine. *Regenerative Medicine*, **5**, 823–835.

Plagnol, A. C., Rowley, E., Martin, P. and Livesey, F. 2009. Industry perceptions of barriers to commercialization of regenerative medicine products in the UK. *Regenerative Medicine*, **4**, 549–559.

Rao, M. S. 2011. Funding translational work in cell-based therapy. *Cell Stem Cell*, **9**, 7–10.

Ratcliffe, E., Thomas, R. J. and Williams, D. J. 2011. Current understanding and challenges in bioprocessing of stem cell-based therapies for regenerative medicine. *British Medical Bulletin*, **100**, 137–155.

Rawe, J. 2000. What will be the 10 hottest jobs? *Time*. http://www.time.com/time/magazine/article/0,9171,997028,00.html

Rayment, E. A. and Williams, D. J. 2010. Concise review: mind the gap: challenges in characterizing and quantifying cell- and tissue-based therapies for clinical translation. *Stem Cells*, **28**, 996–1004.

Reinmuller, B. and Ljungqvist, B. 2003. Modern cleanroom clothing systems: people as a contamination source. *PDA Journal of Pharmaceutical Science and Technology/PDA*, **57**, 114–125.

Rios, M. 2011. Industry educational platforms drive commercialisation objectives. *Bioprocess International*, **9**, 50–53.

Rowley, J. 2010. Developing cell therapy biomanufacturing processes. *Chemical Engineering Progress*, **106**, 50–55.

Serakinci, N., Hoare, S. F., Kassem, M., Atkinson, S. P. and Keith, W. N. 2006. Telomerase promoter reprogramming and interaction with general transcription factors in the human mesenchymal stem cell. *Regenerative Medicine*, **1**, 125–131.

Simonsen, J. L., Rosada, C., Serakinci, N., Justesen, J., Stenderup, K., Rattan, S. I., Jensen, T. G. and Kassem, M. 2002. Telomerase expression extends the proliferative life-span and maintains the osteogenic potential of human bone marrow stromal cells. *Nature Biotechnology*, **20**, 592–596.

Sommerfield, S. and Strube, J. 2005. Challenges in biotechnology production – generic processes and process optimsation for monoclonal antibodies. *Chemical Engineering Progress*, **44**, 1123–1137.

Stenderup, K., Justesen, J., Clausen, C. and Kassem, M. 2003. Aging is associated with decreased maximal life span and accelerated senescence of bone marrow stromal cells. *Bone*, **33**, 919–926.

Sweet, D. J. 2010. Stepping stones on solid ground. *Cell Stem Cell*, **6**, 493.

Thiel, K. A. 2004. Biomanufacturing, from bust to boom…to bubble? *Nature Biotechnology*, **22**, 1365–1372.

Thomas, E. D., Lochte, H. L., JR., Lu, W. C. and Ferrebee, J. W. 1957. Intravenous infusion of bone marrow in patients receiving radiation and chemotherapy. *The New England Journal of Medicine*, **257**, 491–496.

Thomas, R. J., Hourd, P. C. and Williams, D. J. 2008. Application of process quality engineering techniques to improve the understanding of the in vitro processing of stem cells for therapeutic use. *Journal of Biotechnology*, **136**, 148–155.

Titmarsh, D. and Cooper-White, J. 2009. Microbioreactor array for full-factorial analysis of provision of multiple soluble factors in cellular microenvironments. *Biotechnology and Bioengineering*, **104**, 1240–1244.

Trusheim, M. R., Berndt, E. R. and Douglas, F. L. 2007. Stratified medicine: strategic and economic implications of combining drugs and clinical biomarkers. *Nature Reviews. Drug Discovery*, **6**, 287–293.

VA-PBM-Services. 2011. Sipuleucel-T (Provenge). *National Drug Monograph*. USA.

Veraitch, F. S., Scott, R., Wong, J. W., Lye, G. J. and Mason, C. 2008. The impact of manual processing on the expansion and directed differentiation of embryonic stem cells. *Biotechnology and Bioengineering*, **99**, 1216–1229.

Willoughby, N., Martin, P. and Titchener-Hooker, N. 2004. Extreme scale-down of expanded bed adsorption: Purification of an antibody fragment directly from recombinant *E. coli* culture. *Biotechnology and Bioengineering*, **87**, 641–647.

Yeatts, A. B. and Fisher, J. P. 2011. Bone tissue engineering bioreactors: dynamic culture and the influence of shear stress. *Bone*, **48**, 171–181.

Zoro, B. J., Owen, S., Drake, R. A. and Hoare, M. 2008. The impact of process stress on suspended anchorage-dependent mammalian cells as an indicator of likely challenges for regenerative medicines. *Biotechnology and Bioengineering*, **99**, 468–474.

Index

age-related macular degeneration (AMD), 82
alginate hydrogels, 19
Alliance for Regenerative Medicine (ARM), 156
American Society for Testing and Materials (ASTM), 167
Amgen, 221
angiogenesis, 80–1, 83–91
  osteogenesis interaction, 91–2
  co-culture protocol, Plate III
angiogenic growth factors, 86–7
  target cell use in bone formation or remodelling, 88
angiopoeitin (Ang), 88–9
Apligraf, 157
Archimedes principle, 178
arginine-glycine-aspartate motif, 8
ASTM F2027, 112
ASTM F2383, 111
ASTM F2027-08, 112
ASTM F2150-07, 112
ASTM F2210-02, 120
ASTM F2211-04, 110
ASTM F2212-09, 112
ASTM F2315-11, 120
ASTM F2386-04, 120
ASTM F2450-10, 112
ASTM F2451-05, 191
ASTM F2451-10, 120
ASTM F2603-06, 115
ASTM F2664-07, 116
ASTM F2739-08, 119–20
ASTM F2791-09, 115
ASTM F2900-11, 119, 120
attenuated total reflectance (ATR), 173

attenuated total reflectance FTIR (ATR-FTIR), 173

β-fibroblast growth factor (β-FGF), 89
BacT/ALERT, 153
batch reactors, 35
  AT75 cell culture flask, 36
  Techne Flask, 37
Beer-Lambert equation, 199
Bethe equation, 201
Big Pharma, 215
bioactivity, 185, 188
bioactivity tests, 185
biobanking, 145
biocompatibility, 186
biomaterials
  chemical properties, 168–73, 174
    crystallinity of biomaterial, 170–1
    EDS – quantitative analysis of electrospun PLA scaffold, Plate VI
    elemental composition, 171–2
    elemental maps of calcium, phosphorus, oxygen and nitrogen, Plate V
    Fourier transform infrared spectroscopy, 172–3
    molecular weight, 170
    schematic representation of ATR-FTIR operation, 174
    USP chemical tests, 169
  degradation and stability in physiological fluids, 181–4
    different techniques for biomaterials characterisation, 183
    metallic materials degradation, 184

242 Index

biomaterials (*cont.*)
  imaging methods for measuring porosity, 173–8
    abbreviation for methods, 177
    magnetic resonance imaging, 177–8
    optical coherence tomography, 176
    optical microscopy, 176
    pore size characterisation techniques, 175
    scanning electron microscopy, 175
    transmission electron microscopy, 176
    types of pores, 175
    X-ray micro-computed tomography, 176–7
  implant–tissue interface tests, 184–93
    application specific tests, 188–92
    bone tissue implants tests, 186–8
    roughness evaluation, 193
    simulated body fluids composition, 187
    surface morphology/topography, 192–3
  limitations of current standardised testing methods, 193–4
  need for standard testing methods, 166–7
  physical characterisation, 178–9
    density, 178
    nuclear magnetic resonance, 179
    porosity (mercury intrusion), 178–9
  standardised chemical analysis, 167–8
    surface and bulk chemistry analysis, 168
  standardised chemical analysis and testing, 166–94
  surface properties, 179–81
    analytical techniques, 179–80
    hydrated samples surface analysis, 180–1
biopreservation, 145
bioprocessing
  bioreactor configurations, 35–9
  bioreactor culture measurements, 45–6
  bioreactor scaffold materials and architectures, 39–43
  future trends, 48–9

  mass transfer in tissue engineering bioreactor, 43–5
  tissue engineering process design, 46–8
    bioreactor design flowchart, 47
  two- and three-dimensional tissue culture engineering, 34–49
bioreactors
  configurations, 35–9
  culture measurements, 45–6
    physical, chemical and biological parameters, 45
  mass transfer in tissue engineering, 43–5
  scaffold materials and architectures, 39–43
    types, 43
bone, 55, 68–70
  cell culture, 56–60
  process flow for hot embossing, 69
  tissue culture, 60–4
    cell lines, 60–1
    primary culture from isolated tissues, 61–3
    stem cells, 63–4
  tissue culture modelling *in vitro*, 55–6
bone cells, 58–9
  osteoclast cell culture substrate, 59
  peripheral blood mononuclear cells, Plate II
  rat calvarial osteoblasts, Plate I
bone formation, 83–91
bremsstrahlung radiation, 201
British Pharmacopeia (BP), 167
British Standards Institute (BSI), 109
BSI PAS 84-2008, 110

caesium-137, 202
cancellous bone, 55
cartilage, 55, 70
  cell culture, 56–60
  tissue culture, 64–7
    cell lines, 64
    primary culture from isolated tissues, 65–6
    stem cells, 66–7
  tissue culture modelling *in vitro*, 55–6
cartilage cells, 59–60

cell attachment, 41–2
  MG63 cells, 42
cell behaviour, 29
cell biology
  ECM and cell interaction, 9–12
  ECM and mechanical signalling,
    12–13
  extracellular matrix (ECM), 5–9
  future trends, 13
  tissue engineering, 3–14
cell engineering
  standards, 107–23
    characterisation of biomaterials
      and biomolecules, 112
    characterisation of cell-seeded
      scaffolds, 119–20
    characterisation of tissue scaffolds,
      112–19
    characterization of cells and
      cell–surface interactions, 121–2
    lexicon, 109–10
    manufacture, processing and
      storage, 120–1
    national and international
      standards support tissue
      engineering and regenerative
      medicine, 109
    role, 111
    standardisation, 110
cell interaction, 9–12
cell sorting, 156
cell surface modification, 41–2
cell therapy, 214
  cell-based therapy vs monoclonal
    antibody therapies, 230–4
  clinical trial patient numbers: cell
    based vs conventional therapies,
    234
  clinical trial structures and
    manufacturing demand, 234
  current overview of manufacturing
    processes, 231
  cells as therapies, 212–19
    cell therapy industry, 213–14
    commercially available cell therapy
      and regenerative products, 219
    defining cell therapies, regenerative
      medicine and tissue engineering,
      214–15

key challenges in the industry, 214
usage of the terms 'tissue
  engineering,' 'regenerative
  medicine' and 'cell therapy,'
  Plate VII
commercial manufacture, 212–35
considerations for translation from
  laboratory bench to factory floor,
  221–4
  development rates of autologous vs
    allogeneic products, 223–4
  intellectual property, 222
  legacy products, 222
  process robustness and clinical
    success, 222–3
current state of the industry, 215–19
  autologous vs allogeneic cell
    therapy, 215–18
  current CTI products and those in
    the pipeline, 220
  potential advantages and
    challenges of autologous and
    allogeneic cell therapies, 217–18
  testing challenges in cell therapy
    manufacture, 216
  the 'Cost Iceberg,' 228
emergence of cell therapy platform
  manufacturing process, 230–4
  'An Emerging Cell Therapy
    Platform Process,' 232
  cost, 232
  practical and logistical advantages,
    233
  process, 233
  the negatives, 233–4
essential steps needed in translation
  and commercialisation, 229
key regulatory requirements for
  commercial manufacture, 225–8,
  229–30
  essential steps towards a
    commercial process, 228
  extended culture duration, 226
  process vs product safety, 227–8
  product safety, 227
  reproducibility and comparability,
    225
  serum elimination, 225–6
quality control, 148–62

244　Index

cell therapy (*cont.*)
　commercial quality control/
　　quality assurance in large-scale
　　manufacture, 157, 159–62
　ensuring well-defined cell therapy
　　products, 151–7, 158
　QC in cell therapy manufacture,
　　148–9
　requirements for allogeneic vs
　　autologous cell-based therapies,
　　150
　testing challenges in cell therapy
　　manufacture, 150
　transition from laboratory to
　　commercial-scale manufacture,
　　220–4
　process development, 224
cell therapy industry (CTI), 213
cell–cell batch variability, 117
cellular interfaces, 28
chemical analysis
　standardised for biomaterials, 166–94
chondrocyte, 59–60
cobalt-60, 202, 203
collagen, 6–72
　controlling mechanical properties,
　　21–6
　　cross-linking, 25
　　density, 21–3
　　fibril diameter, 23–4
　　packing and orientation, 24–5
　elements of tissue complexity, 26–30
　　cellular interfaces, 28
　　glass fibres embedded into collagen
　　　hydrogels, 27
　　gradients in 3D matrices, 28–30
　engineering cell- and matrix-rich
　　tissues using collagen scaffolds,
　　18–20
　future trends, 30–1
　three-dimensional biomatrix
　　development and control, 18–31
collagen cross-linking, 25
collagen density, 21–3
　PC gel application and configuration,
　　22
collagen fibril diameter, 23–4
collagen hydrogel, 20

collagen orientation, 24–5
collagen packing, 24–5
collagen scaffolds
　engineering cell- and matrix-rich
　　tissues, 18–20
　collagen hydrogel set-up, 20
collagen type I scaffolds, 19
colony-forming unit (CFU), 205
Committee for European
　Standardization, 109
compact bone, 55
Compton scattering, 199
continuous stirred tank reactor
　(CSTR), 35
Coomassie Blue, 169
cortical bone, 55

Darcy permeability coefficient, 114
decorin, 23–4
dedifferentiated chondrocyte, 65
　expansion, 65–6
　re-differentiation, 66
density, 175, 178
differential scanning calorimetry
　(DSC), 171
DIN 4768, 193
dose mapping, 207
dynamic vapour sorption (DVS), 171
Dynamitron, 203

E-governance, 129–30
electron probe X-ray microanalysis
　(EDX), 171
electron spectroscopy for chemical
　analysis (ESCA), 179
ellipsometry, 181
embryonic stem cells (ESCs), 63, 66
embryonic vasculogenesis, 79–81
Enbrel (Etanercept), 221
energy dispersive spectroscopy (EDS),
　171, 172
enzyme-linked immunosorbent assay
　(ELISA), 170
epigenetic profiling, 156
EU Tissues and Cells Directive, 131–2
European Centre for the Validation of
　Alternative Methods (ECVAM),
　130

© Woodhead Publishing Limited, 2013

European Commission Medical
  Directive 93/42/EEC, 108, 110
extracellular matrix (ECM), 4–5, 5–9,
  9–12, 12–13, 14
  collagen, 6–7
  fibronectin, 7–8
  glycosaminoglycans, 9
  laminin, 8–9

fibronectin, 7–8
flat bed reactors, 36
  microreactor, 38
flow cytometry, 156
fluidised bed bioreactors,
  37–8
  packed bed and fluidised bed
    bioreactors, 39
foetal calf serum (FCS), 153, 225–6
Fourier transform infrared (FTIR)
  spectroscopy, 172–3, 179
Fourier transform infrared microscopy,
  173

gas flow porometry, 179
gene expression, 156
Genzyme, 153
glycosaminoglycans (GAG), 9
good clinical practice (GCP), 129
good laboratory practice (GLP)
  application to human cell culture
    systems, 143–5
    biobanking and biopreservation,
      145
    isolation of stem cells from bone
      marrow, 143–5
  characterisation, 135–8
    test items, 138
    test system, 135–8
  governing bodies, 129–32
  in vitro cell culture applications,
    127–45
  overview, 127–9
    definition, 128
    history, 127–8
    industries, 129
  research monitoring and quality
    assurance personnel, 142–3
  resources for compliance, 132–5

facilities and equipment, 133–5
organized structure, 132
personnel and personnel
  management, 132–3
study director, 133
results documentation, 141–2
  final report and archiving, 142
  raw data, 141–2
standards and regulations, 138–41
  protocol, 139–40
good manufacturing practice (GMP),
  129
gradients, 28–30
  O_2 gradient formation within a cell-
    seeded 3D collagen construct,
    30
growth factor signalling, 11

H1 human embryonic stem cells (H1
  HESC), 64
hard tissue
  two- and three-dimensional
    culture methods for tissue
    engineering, 54–71
    bone and cartilage cells, 56–60
    bone tissue culture, 60–4
    cartilage tissue culture, 64–7
    overview, 54–6
heparan sulphate proteoglycans
  (HSPG), 9
high-energy electron
  sources, 203
high-energy electrons, 201–2
high-energy photons,
  199–201
high-resolution electron
  energy loss (HREELS), 179
human chondrosarcoma cell line, 64
human immortalised chondrocyte cell
  line, 64
human osteosarcoma cell lines, 60
Human Tissue Authority (HTA),
  145
hypoxia, 89–90

I-Ax Technologies Inc., 204
Image J software, 174
immersion test, 184

© Woodhead Publishing Limited, 2013

246 Index

implant–tissue interface tests, 184–93
  application specific tests, 188–92
    implant–blood interface testing, 189
    implant–bone tissue interface testing, 188–9
    implant materials for cartilage regeneration, 191–2
    implant materials for ligament regeneration, 190–1
    implant materials for pulmonary delivery, 189–90
  bone tissue implants tests, 186–8
  simulated body fluids composition, 187
  surface morphology/topography, 192–3
  roughness evaluation, 193
in vitro cell culture
  good laboratory practice (GLP), 127–45
    application to human cell culture systems, 143–5
    characterisation, 135–8
    governing bodies, 129–32
    overview, 127–9
    research monitoring and quality assurance personnel, 142–3
    resources for compliance, 132–5
    results documentation, 141–2
    standards and regulations, 138–41
in vitro prevascularisation, 92–3
  cell proliferation, 93
  co-cultures of human osteoblast cells (HOBs) and human umbilical vein endothelial cells (HUVECs), Plate IV
insulated core transformer, 203
integrins, 9–10
Interagency Coordinating Committee on the Validation of Alternative Methods (ICCVAM), 130–1
International Organisation for Standardisation (ISO), 109, 167
intrinsic viscosity (IV), 170
ionising radiation
  interaction with matter, 198–202
  high-energy electrons, 201–2

high-energy photons, 199–201
ionisation of molecules, 198–9
mass attenuation coefficients for water for high-energy photons, 200
penetration of 10 MeV electrons through water, 202
penetration of 1.25 MeV photons through water, 200
sources, 202–4
commercial radiation sources, 203–4
gamma radiation and X-ray sources, 202–3
high-energy electron sources, 203
validation and international standards, 204–9
international standards, 205–9
microbial standard distribution of resistance, 206
principles, 204–5
ISO 10993, 111
ISO 4287, 115, 193
ISO 4288, 193
ISO 9000, 108, 109
ISO 10993, 118, 184, 182, 189, 191
ISO 11137, Parts 1–3, 2006, 206
ISO 13408, 120
ISO 21534, 190
ISO/TS 10993, 192

laminin, 8–9
laser scanning confocal microscopy (LSCM), 176
linacs, 203
Lonza Biologics, 161

magnetic resonance imaging (MRI), 177–8
master cell bank (MCB), 161
matrix biology
  ECM and cell interaction, 9–12
  ECM and mechanical signalling, 12–13
  extracellular matrix (ECM), 5–9
  future trends, 13
  tissue engineering, 3–14

Index   247

mechanical signalling, 12–13
Medicines and Healthcare products
    Regulatory Agency (MHRA),
    110
membrane bioreactors, 38–9
    modules, 40
mesenchymal stem cells (MSCs), 57,
    66, 226
microvascular networks, 79–81
    microvasculature formation, 80
MODA paperless QC system, 161
molecular weight (MW), 170
monoclonal antibody (mAb), 221
mouse osteocyte cell line, 60
murine chondrocyte cell line, 64
murine osteoblastic cells lines, 60
myocardial infarction, 81–2

National Toxicology Program
    Interagency Centre for the
    Evaluation of Alternative
    Toxicological Methods
    (NICEATM), 131
natural polymers, 40
near X-ray absorption fine structure
    (NEXAFS), 180
nuclear magnetic resonance (NMR),
    179

optical coherence tomography, 176
optical microscopy, 176
Osiris Therapeutics, 157
osteoblasts, 58, 61
osteoclasts, 62–3
osteocytes, 55, 58, 61
osteogenesis
    angiogenesis interaction, 91–2
        co-culture protocol, Plate III
osteoprogenitors, 58, 61–2

packed bed bioreactors, 36
photon in/photon out techniques, 181
plastically compress (PC), 22
platelet-derived growth factor (PDGF),
    88–9
polyethylene glycol (PEG), 19
polymerase chain reaction, 137
porosimetry, 175
porosity (mercury intrusion), 178–9

potentiodynamic test, 184
Prochymal, 157
proteases, 11–12
Provenge (Dendreon), 215, 222, 227,
    234

quality assurance, 149
quality control, 149
    cell and tissue engineering, 148–62
    cell therapy manufacture, 148–9
    commercial quality control/quality
        assurance in large-scale
        manufacture, 157, 159–62
        cell therapy platform process, 159
        key considerations in production
            of typical autologous cell-based
            therapies, 160
    defining critical quality attributes for
        cell therapies, 151–7
        identity, 156
        potency, 156–7
        purity, 154–6
        safety, 152–4
    ensuring well-defined cell therapy
        product, 151–7, 158
        stability matrix, 158
    requirements for allogeneic vs
        autologous cell-based therapies,
        150
    testing challenges in cell therapy
        manufacture, 150

radiation dose, 207
Raman sum frequency generation
    (SFG), 179
RAW 264.7, 60
regenerative medicine, 107, 214
Rhodotron, 203
Rhodotron TT-1000 system, 204

sample item portion (SIP), 207
scanning electron microscopy (SEM),
    175
scanning probe microscopy (SPM),
    179–80
SDS-polyacrylamide gel electrophoresis
    (SDS-PAGE), 170
second-generation products,
    230

© Woodhead Publishing Limited, 2013

248 Index

secondary ion mass spectrometry
(SIMS), 179
simulated body fluid (SBF), 186–8
soft tissue
two- and three-dimensional tissue
culture bioprocessing for tissue
engineering, 34–49
bioreactor configurations, 35–9
bioreactor culture measurements,
45–6
bioreactor scaffold materials and
architectures, 39–43
future trends, 48–9
mass transfer in bioreactor, 43–5
process design, 46–8
standard distribution of resistance
(SDR), 205
standard operating procedure (SOP),
140–1
standards
cell and tissue engineering, 107–23
characterisation of biomaterials and
biomolecules, 112
characterisation of cell-seeded
scaffolds, 119–20
characterisation of tissue scaffolds,
112–19
blind-end, through and closed
pores, 114
key factors in characterising
hydrogels, 119
methods, 113
porosity structures, 116
characterisation of cells and cell–
surface interactions, 121–2
lexicon, 109–10
manufacture, processing and storage,
120–1
national and international standards
support tissue engineering and
regenerative medicine, 109
role, 111
standardisation, 110
static time-of-flight secondary ion mass
spectrometry (static ToF SIMS),
180
stem cells, 63–4
human mesenchymal stem cells
(hMSCs), 63

isolation from bone marrow,
143–5
events applied to a cell culture
system, 144
sterilisation procedures, 197–210
future trends, 210
interaction of ionising radiation with
matter, 198–202
high-energy electrons, 201–2
high-energy photons, 199–201
ionisation of molecules, 198–9
mass attenuation coefficients for
water for high-energy photons,
200
penetration of 10 MeV electrons
through water, 202
penetration of 1.25 MeV photons
through water, 200
sources of ionising radiation, 202–4
commercial radiation sources,
203–4
gamma radiation and X-ray
sources, 202–3
high-energy electron
sources, 203
validation and international
standards of sterilisation by
ionising radiation, 204–9
microbial standard distribution of
resistance, 206
principles, 204–5
sterility assurance level (SAL), 198
stopping power, 201
sum frequency generation
(SFG), 181
surface plasmon resonance (SPR),
179, 181
synthetic polymers, 40

teratoma, 227
test items, 138
test system, 135–8
Testing Laboratory Registration Act,
127
thin film X-ray diffraction (TF-XRD),
172, 188
tissue allografts
sterilisation procedures, 197–210
future trends, 210

© Woodhead Publishing Limited, 2013

Index 249

interaction of ionising radiation with matter, 198–202
sources of ionising radiation, 202–4
validation and international standards of sterilisation by ionising radiation, 204–9
tissue culture
bone and cartilage cells, 56–60
cell characterisation, 58–60
cell isolation, 57
media formulations, 57–8
tissue sources for cell harvesting, 56–7
bone tissue, 60–4
cartilage tissue, 64–7
overview, 54–6
two- and three-dimensional methods for hard tissue engineering, 54–71
two- and three-dimensional soft tissue bioprocessing for tissue engineering, 34–49
two-and-a-half and three-dimensional methods for hard tissues, 68–71
two-dimensional methods for hard tissues, 67–8
process flow for soft lithography, 68
tissue-engineered constructs vascularisation, 77–96
angiogenesis and bone formation, 83–91
angiogenesis and osteogenesis interaction, 91–2
angiogenic diseases, 81–3
embryonic vasculogenesis, 79–81
in vitro prevascularisation, 92–3
tubular formation, 94
vascular tissue engineering cell sources, 91
tissue-engineered medical products (TEMP), 108
tissue engineering, 107, 214
cell and matrix biology, 3–14
ECM and cell interaction, 9–12
ECM and mechanical signalling, 12–13
extracellular matrix (ECM), 5–9

future trends, 13
standards, 107–23
characterisation of biomaterials and biomolecules, 112
characterisation of cell-seeded scaffolds, 119–20
characterisation of tissue scaffolds, 112–19
characterisation of cells and cell–surface interactions, 121–2
lexicon, 109–10
manufacture, processing and storage, 120–1
national and international standards support tissue engineering and regenerative medicine, 109
role, 111
standardisation, 110
two- and three-dimensional hard tissue culture methods, 54–71
bone and cartilage cells, 56–60
bone tissue culture, 60–4
cartilage tissue culture, 64–7
overview, 54–6
two- and three-dimensional tissue culture bioprocessing, 34–49
bioreactor configurations, 35–9
bioreactor culture measurements, 45–6
bioreactor scaffold materials and architectures, 39–43
future trends, 48–9
mass transfer in bioreactor, 43–5
process design, 46–8
topography, 192
trabecular bone, 55
transforming growth factor-β, 88–9
transmissible spongiform encephalopathies (TSE), 153
transmission electron microscopy (TEM), 176
tumour growth, 82–3
type I collagen, 7, 18–19, 21, 30–1
type III collagen, 7
type IV collagen, 7
type IX collagen, 23–4
type V collagen, 23–4
type VI collagen, 7

© Woodhead Publishing Limited, 2013

250 Index

ultra scale-down (USD), 224
United States Pharmacopeia (USP), 167

vascular endothelial growth factor (VEGF), 87–9
vascularisation
  angiogenesis and bone formation, 83–91
    HUVEC cells growth, 87
    scaffold design, 84–6
    tubular formation, 85
    vascular tissue engineering, 83–4
  angiogenesis and osteogenesis interaction, 91–2
  angiogenic diseases, 81–3
  embryonic vasculogenesis, 79–81
  in vitro prevascularisation, 92–3
  tissue-engineered constructs, 77–96

tubular formation, 94
  interconnected vessel networks, 96
  schematic diagram, 95
  vascular tissue engineering cell sources, 91
vasculogenesis, 80
VD_{max} method, 207
VIPCO business model, 232

wavelength-dispersive X-ray spectroscopy (WDS), 171, 172

X-ray diffraction (XRD), 170, 171
X-ray micro-computed tomography (MicroCT), 176–7
X-ray photoelectron spectroscopy (XPS), 172, 180, 188

zeta potential measurement, 179

CPSIA information can be obtained at www.ICGtesting.com
Printed in the USA
BVOW11*0052070214

343693BV00001BB/3/P